New Frontiers in Acrylamide Study in Foods

New Frontiers in Acrylamide Study in Foods: Formation, Analysis and Exposure Assessment

Editors

Marta Mesías
Cristina Delgado-Andrade
Francisco J. Morales

MDPI • Basel • Beijing • Wuhan • Barcelona • Belgrade • Manchester • Tokyo • Cluj • Tianjin

Editors
Marta Mesías
Spanish National Research Council (CSIC)
Spain

Cristina Delgado-Andrade
Spanish National Research Council (CSIC)
Spain

Francisco J. Morales
Spanish National Research Council (CSIC)
Spain

Editorial Office
MDPI
St. Alban-Anlage 66
4052 Basel, Switzerland

This is a reprint of articles from the Special Issue published online in the open access journal *Foods* (ISSN 2304-8158) (available at: https://www.mdpi.com/journal/foods/special_issues/Acrylamide_Foods_Formation_Analysis_Exposure).

For citation purposes, cite each article independently as indicated on the article page online and as indicated below:

LastName, A.A.; LastName, B.B.; LastName, C.C. Article Title. *Journal Name* **Year**, *Volume Number*, Page Range.

ISBN 978-3-0365-0030-0 (Hbk)
ISBN 978-3-0365-0031-7 (PDF)

Cover image courtesy of Marta Mesías, Cristina Delgado-Andrade andFrancisco J. Morales

Contents

About the Editors

Marta Mesías graduated in Pharmacy from the University of Seville (Spain) and graduated in Food Science and Technology from the University of Granada (Spain). She is part of the research team studying chemical modifications in processed foods at the Institute of Food Science, Technology and Nutrition (ICTAN). Particularly, her investigations focus on the Maillard reaction, evaluating the technological, nutritional, and toxicological consequences of the appearance of Maillard reaction products during food processing. She combines her research with dissemination, promoting the transfer of knowledge to society and consumer education in terms of food safety.

Cristina Delgado-Andrade is a tenured scientist from the Spanish National Research Council working at the field of Food Science and Health–Diet interactions since 1997. Her major interests are in the exposition to different food processing contaminants and the study of their connection with the progress and development of associated diseases. Thanks to a multidisciplinary training acquired in Spanish and European laboratories, she is currently involved in various scientific projects in collaboration with notable international researchers and institutions. She has also carried out advisory works for the food industry as well as tailored projects according to business necessities.

Francisco J. Morales graduated in Biochemistry and Molecular Biology and with a PhD in Food Chemistry from the Autonomous University of Madrid (Spain). He has been a permanent scientific research staff member at the Spanish National Council for Scientific Research (CSIC) since 1998. He is currently a member of the scientific committee of the Spanish Agency for Food Safety at the European Committee for Standardization (CEN), and at the Commission of the Global Area LIFE at CSIC. He is Director of the research team on chemical modifications in processed foods at the Institute of Food Science, Technology and Nutrition (ICTAN). He served as Researcher at the University of Wageningen (NL) for 2 years and, later, Deputy Scientific Director at the Institute of Science and Technology of Food and Nutrition (ICTAN) from 2011 to 2017. His research covers the study of the beneficial/adverse properties of processed foods from a risk/benefit perspective. He coordinates a multidisciplinary team with complementary skill in food chemistry, food technology, nutrition, human health, and social sciences. Particularly, his investigations focus on the Maillard reaction, covering the technological, nutritional, and toxicological repercussion of Maillard reaction products. He has participated and led a large number of regional, national, and European research projects in addition to contracts for development and innovation with the agro-food private sector. He has transferred a patent under international exploitation, and also acts as supervisor of visiting researchers, predoctoral, students, and technicians. He has published more than 150 scientific articles in peer-reviewed international journals (corresponding to a h-index of 41), and 18 book chapters in the field of quality and food safety.

Editorial

Introduction to the Special Issue: New Frontiers in Acrylamide Study in Foods—Formation, Analysis and Exposure Assessment

Cristina Delgado-Andrade, Marta Mesías * and Francisco J. Morales

Institute of Food Science, Technology and Nutrition (ICTAN), Spanish National Research Council (CSIC), E-28040 Madrid, Spain; cdelgado@ictan.csic.es (C.D.-A.); fjmorales@ictan.csic.es (F.J.M.)
* Correspondence: mmesias@ictan.csci.es; Tel.: +34-91-5492300

Received: 28 September 2020; Accepted: 16 October 2020; Published: 21 October 2020

Abstract: Acrylamide is a chemical contaminant that naturally originates during the thermal processing of many foods. Since 2002, worldwide institutions with competencies in food safety have promoted activities aimed at updating knowledge for a revaluation of the risk assessment of this process contaminant. The European Food Safety Authority (EFSA) ruled in 2015 that the presence of acrylamide in foods increases the risk of developing cancer in any age group of the population. Commission Regulation (EU) 2017/2158 establishes recommended mitigation measures for the food industry and reference levels to reduce the presence of acrylamide in foods and, consequently, its harmful effects on the population. This Special Issue explores recent advances on acrylamide in foods, including a novel insight on its chemistry of formation and elimination, effective mitigation strategies, conventional and innovative monitoring techniques, risk/benefit approaches and exposure assessment, in order to enhance our understanding for this process contaminant and its dietary exposure.

Keywords: acrylamide; chemical process contaminants; Maillard reactions; food safety; risk/benefits; mitigations; exposure

---

Chemical process contaminants are substances formed when foods undergo chemical changes during processing, including heat treatment, fermentation, smoking, drying and refining. Although necessary for making food edible and digestible, heat treatment can have undesired consequences leading to the formation of heat-induced contaminants such as acrylamide. It is well-established that acrylamide is formed when foods containing free asparagine and reducing sugars are cooked at temperatures above 120 °C in low moisture conditions. It is mainly formed in baked or fried carbohydrate-rich foods, as the relevant raw materials contain its precursors. These include cereals, potatoes and coffee beans. In 1994, acrylamide was classified by the International Agency for Research on Cancer as being probably carcinogenic to humans (group 2A), and in 2015, the European Food Safety Authority (EFSA) confirmed that the presence of acrylamide in foods is a public health concern, requiring continued efforts to reduce its exposure.

This special edition assembled nine quality papers, one review and eight research papers, focusing on several acrylamide-related issues, from raw materials to consumer exposure.

Different approaches for acrylamide determination in foods have been critically reviewed by Pan et al. [1], including conventional instrumental analysis methods and the new rapid immunoassay and sensor detection procedures. Advantages and disadvantages of different analysis technologies are compared in order to provide new ideas for the development of more efficient and practical analysis methods and detection equipment. Fernandes et al. [2] set up a high-resolution orbitrap mass spectrometry method for acrylamide measurement, with good repeatability, limit of detection and quantification, as well as enhanced detection sensitivity.

Some of the papers included have focused on the importance of precursor levels in the raw matter and the processing conditions on acrylamide formation. In this sense, Sun and colleagues [3] investigated the effects of nitrogen rate and storage time on potato glucose concentrations in different cultivars, analyzing the relationships between acrylamide, glucose, and asparagine for new cultivars. Mesias et al. [4] evaluated browning, antioxidant capacity and the formation of acrylamide and other heat-induced compound at different stages during the production of block panela (non-centrifugal cane sugar), establishing the juice concentration step as the critical point to settle mitigation strategies. Lee and co-workers [5] assessed the effects of thawing and frying methods on the formation of acrylamide and polycyclic aromatic hydrocarbons (PAHs) in chicken meat. They conclude that air frying could reduce the formation of acrylamide and PAHs in this food matrix at in comparison with deep-fat frying. In the case of cereal-derived products, Fernandes et al. [2] compared the acrylamide levels of biscuits with several production parameters, such as time/cooking temperature, placement on the cooking conveyor belt, color, and moisture. They state that the composition of the raw materials is the most important factor in the acrylamide content; therefore, establishing the level of precursor of ingredients strongly would contribute to the establishment of effective mitigation strategies. Industrial strategies to reduce acrylamide formation in Californian-style green ripe olives were studied by Martín-Veltedor et al. [6], with interesting results for the table olive industry to identify critical points in the production of this type of olives, thus helping to control acrylamide formation in this foodstuff.

It is well-known that potato- and cereal-derived products as well as coffee are important acrylamide sources in the Western diet. The food industry is especially interested in prospective studies dealing with the presence of acrylamide in these elaborations and its evolution in recent years. The study by Mesias et al. [7] evaluated acrylamide levels in seventy potato crisp samples commercialized in Spain with the purpose of updating knowledge about the global situation in this snack sector and evaluate the effectiveness of mitigation strategies applied, especially since the publication of the 2017/2158 Regulation. Results demonstrated that average acrylamide content in 2019 was 55.3% lower compared to 2004, 10.3% lower compared to 2008 and very similar to results from 2014, evidencing the effectiveness of mitigation measures implemented by Spanish potato crisp manufacturers. However, 27% of samples exhibited concentrations above the benchmark level established in the Regulation, which suggests that efforts to reduce acrylamide formation in this sector must continue. The same research team also developed a survey in 730 Spanish households to identify culinary practices which might influence acrylamide formation during the domestic preparation of French fries and their compliance with the acrylamide mitigation strategies described in the same document [8]. They conclude that although habits of the Spanish population are in line with recommendations to mitigate acrylamide during French fry preparation, educational initiatives disseminated among consumers would reduce the formation of this contaminant and, consequently, exposure to it in a domestic setting. Finally, an assessment of healthy and harmful Maillard reaction products (melanoidins and acrylamide) in a sun-dried coffee cascara beverage was developed by Iriondo-DeHond et al. [9], analyzing its safety and health-promoting properties. The novel beverage is proposed as a potential sustainable alternative for instant coffee, with low caffeine and acrylamide levels and a healthy composition of nutrients and antioxidants.

We hope that this Special Issue will be interesting for researchers engaged in the acrylamide issue in foods, including a novel insight on its chemistry of formation and elimination, effective mitigation strategies, classical and novel monitoring techniques, risk/benefit approaches, and exposure assessment, in order to enhance our understanding for this process contaminant and its dietary exposure.

Conflicts of Interest: The authors declare no conflict of interest.

References

1. Pan, M.; Liu, K.; Yang, J.; Hong, L.; Xie, X.; Wang, S. Review of Research into the Determination of Acrylamide in Foods. *Foods* **2020**, *9*, 524. [CrossRef] [PubMed]

2. Fernandes, C.; Carvalho, D.; Guido, L. Determination of Acrylamide in Biscuits by High-Resolution Orbitrap Mass Spectrometry: A Novel Application. *Foods* **2019**, *8*, 597. [CrossRef] [PubMed]
3. Sun, N.; Wang, Y.; Gupta, S.K.; Rosen, C.J. Potato Tuber Chemical Properties in Storage as Affected by Cultivar and Nitrogen Rate: Implications for Acrylamide Formation. *Foods* **2020**, *9*, 352. [CrossRef]
4. Mesias, M.; Delgado-Andrade, C.; Gómez-Narváez, F.; Contreras-Calderón, J.; Morales, F. Formation of Acrylamide and other Heat-Induced Compounds during Panela Production. *Foods* **2020**, *9*, 531. [CrossRef]
5. Lee, J.; Han, J.; Jung, M.; Lee, K.; Chung, M. Effects of Thawing and Frying Methods on the Formation of Acrylamide and Polycyclic Aromatic Hydrocarbons in Chicken Meat. *Foods* **2020**, *9*, 573.
6. Martín-Vertedor, D.; Fernández, A.; Mesías, M.; Martínez, M.; Díaz, M.; Martín-Tornero, E. Industrial Strategies to Reduce Acrylamide Formation in Californian-Style Green Ripe Olives. *Foods* **2020**, *9*, 1202. [CrossRef]
7. Mesias, M.; Nouali, A.; Delgado-Andrade, C.; Morales, F. How Far is the Spanish Snack Sector from Meeting the Acrylamide Regulation 2017/2158? *Foods* **2020**, *9*, 247. [CrossRef] [PubMed]
8. Mesias, M.; Delgado-Andrade, C.; Morales, F. Are Household Potato Frying Habits Suitable for Preventing Acrylamide Exposure? *Foods* **2020**, *9*, 799. [CrossRef]
9. Iriondo-DeHond, A.; Elizondo, A.; Iriondo-DeHond, M.; Ríos, M.; Mufari, R.; Mendiola, J.; Ibañez, E.; del Castillo, M. Assessment of Healthy and Harmful Maillard Reaction Products in a Novel Coffee Cascara Beverage: Melanoidins and Acrylamide. *Foods* **2020**, *9*, 620. [CrossRef]

Publisher's Note: MDPI stays neutral with regard to jurisdictional claims in published maps and institutional affiliations.

Review

Review of Research into the Determination of Acrylamide in Foods

Mingfei Pan [1,2]**, Kaixin Liu** [1,2]**, Jingying Yang** [1,2]**, Liping Hong** [1,2]**, Xiaoqian Xie** [1,2] **and Shuo Wang** [1,2,*]

[1] State Key Laboratory of Food Nutrition and Safety, Tianjin University of Science & Technology, Tianjin 300457, China; panmf2012@tust.edu.cn (M.P.); lkx13642168374@163.com (K.L.); yangjy0823@126.com (J.Y.); honglpstu@163.com (L.H.); xiexiaoqian8135@163.com (X.X.)

[2] Key Laboratory of Food Nutrition and Safety, Ministry of Education of China, Tianjin University of Science and Technology, Tianjin 300457, China

* Correspondence: s.wang@tust.edu.cn; Tel.: +86-022-60912493

Received: 30 March 2020; Accepted: 20 April 2020; Published: 22 April 2020

Abstract: Acrylamide (AA) is produced by high-temperature processing of high carbohydrate foods, such as frying and baking, and has been proved to be carcinogenic. Because of its potential carcinogenicity, it is very important to detect the content of AA in foods. In this paper, the conventional instrumental analysis methods of AA in food and the new rapid immunoassay and sensor detection are reviewed, and the advantages and disadvantages of various analysis technologies are compared, in order to provide new ideas for the development of more efficient and practical analysis methods and detection equipment.

Keywords: acrylamide; detection; rapid methods; food safety

1. Introduction

Acrylamide (AA) is a small molecule organic compound that exists in solid form at normal temperature and pressure. It is sensitive to the light and can be initiated to polymerize to form polyacrylamide under ultraviolet conditions [1,2]. Therefore, AA is a commonly-used polymerization monomer in industry. In 1994, AA was classified as a "probable carcinogen" by the International Cancer Agency (IARC) and in April 2002, researchers demonstrated that plant foods rich in carbohydrates and low in protein are prone to produce large amounts of AA during high-temperature (>120 °C) processing such as frying and baking [3,4]. This result has caused widespread concern about this compound worldwide. The thermal processing of food is an indispensable process in modern food processing. Under heat treatment such as frying and baking, foods rich in starch and other carbohydrates have color, flavor, and other characteristics added through the Maillard reaction, which is the main way to form AA [5–7]. Recently, AA generation has been associated with high sterilization temperatures, mainly involving the formation of AA in fat-rich foods such as ripe black table olives [8,9]. The content of AA in high-carbohydrate foods with different thermal processing methods is different, and within a certain temperature range, the content of AA increases with the processing time and temperature [10–13]. According to the report of Commission Regulation (EU) 2017/2158 which established mitigation measures and benchmark levels for the reduction of the presence of acrylamide in food, the average AA content in food (processed cereal products, coffee substitutes, etc.) is in the range of 40–4000 µg kg^{-1} [14]. Since foods proposed to monitor the presence of AA in the Commission Regulation (EU) 2019/1888 (potato products, bakery products, cereal products, and others such as dried fruits, olives in brine) are an important part of human food, it is particularly important to deepen the research and quantitative analysis of the process control of AA content in foods [15–17].

In recent years, the mechanism of AA production and its mutagenesis and carcinogenesis in the human body have been gradually revealed [18–21] and related strategies for AA detection in various foods have been successively developed. These strategies are not only used for the analysis of AA content in foods, but also provide a reliable judgement of AA risk level [22]. It has been reported that the tolerable daily intake (TDI) of neurotoxic and carcinogenic AA is 40 and 2.6 $\mu g\ kg^{-1}\ day^{-1}$, respectively [23]. On the other hand, the matrix of heat-processed foods rich in carbohydrates is usually complicated. In addition, AA has a small molecular weight (Mr = 71.08 $g\ mol^{-1}$), high reactivity, and other characteristics, which makes it difficult to perform accurate quantitative analysis of AA. Therefore, it is of great significance to develop accurate, sensitive, and anti-interference methods for the analysis and detection of AA content in foods.

This paper reviews the conventional instrumental methods for AA detection in foods and new types of analytical methods such as rapid immunoassays, supramolecular recognition, and nano-biosensors, and comprehensively evaluates the advantages and the shortcomings of various analytical techniques, aiming to provide new ideas for the development of more-efficient and practical analytical methods and testing devices, so as to provide technical support for the detection and risk assessment of AA in foods.

2. Instrumental Analysis Strategies for AA Content in Foods

Up to now, instrumental analysis based on the principles of chromatography and mass spectrometry including high performance liquid chromatography (HPLC) [24–26], gas chromatography (GC) [27–29], liquid chromatography tandem mass spectrometry (LC-MS/MS) [30–32], and gas chromatography-mass spectrometry (GC-MS) [33] have still been the main methods to detect AA content in foods. With high accuracy and sensitivity, as well as good stability and reproducibility, these kinds of methods are the most reliable for analysis and detection of AA. Therefore, although these kinds of methods need expensive equipment and are high in detection cost, they are still the main methods for detecting AA content in food. Luo et al. have developed a non-aqueous reaction system based on the GC-MS method for rapid and sensitive detection of AA in food matrices [34]. Under mild reaction conditions (40 °C), concentrated AA can complete the reaction with flavanol in 1 min, which simplifies the derivatization reaction process and improves the stability of the detection results. Under optimal conditions, this developed GC-MS method has a linear response range of 0.005–4 $\mu g\ mL^{-1}$ with correlation coefficient (R^2) at 0.99993 in food matrices. The limit of detection (LOD, S/N = 3) and the relative standard deviation (RSD, n = 6) are achieved at 0.7 $\mu g\ kg^{-1}$ and 2.3–6.1%, respectively, showing good accuracy, sensitivity, and repeatability, which can meet the needs of detection of AA in food matrix. However, due to the high polarity, low volatility, and low molecular weight of AA, the derivatization process is often needed to enhance the stability of AA, and further improve the detection sensitivity of GC and its combination technology. LC-MS/MS, however, has no derivatization process, greatly reducing the detection time and meeting the requirements for a green environment [35,36]. Calbiani and his co-workers established a fast and accurate method for the determination of AA in cooked food samples by reversed-phase LC-MS coupled with electrospray [37]. An acidified water extraction step without purification was used in this method, simplifying sample-processing procedures. Remarkable results (LOD: <15 $\mu g\ kg^{-1}$; LOQ: <25 $\mu g\ kg^{-1}$) were obtained for intraday repeatability (RSD < 1.5%) and between-day precision (RSD < 5%), demonstrating that this method is suitable for the determination of AA in cooked food products. Galuch et al. extracted AA from coffee samples by the method of dispersion liquid–liquid microextraction, combined with ultra-performance LC-MS/MS and standard addition method, obtaining good detection sensitivity (LOD: 0.9 $\mu g\ L^{-1}$; limit of quantitation (LOQ): 3.0 $\mu g\ L^{-1}$) and precision (internal and inter-assay precision: 6–9%) [38]. Tolgyesi developed a hydrophilic interaction liquid chromatography tandem mass spectrometric (HILIC-MS/MS) to determine AA in gingerbread samples with high sugar content [39]. The proposed method had acceptable accuracy (101–105%) and precision (2.9–7.6%) with a LOQ of 20 $\mu g\ kg^{-1}$. At the same time, the method was also applied to other food samples (bread, roasted coffee, instant coffee, cappuccino powder, and fried potatoes), and the tested AA content was lower than the EU-set level. Additionally, because of the good separation effect, LC-MS/MS can also be applied in simultaneous detection of AA and other harmful substances in one sample, which has good application value [40,41]. Wu et al. used isotope-dilution ultra-performance LC-MS/MS for simultaneous detection of 4-methylimidazole and AA in 17 commercial biscuit products [42], revealing the wide presence of 4-methylimidazole and AA in biscuit products. This method was validated with respect to linearity, LOQ, precision, trueness, and measurement uncertainty and offers a reliable and sensitive tool for 4-MI and AA measurements in biscuit products.

In addition, because foods are complex matrices, analytical methods using large precision instruments often require a relatively tedious process for sample purification. Therefore, developing effective and reliable materials for sample pretreatment and purification is meaningful to improve sensitivity and accuracy of AA detection, and has very important application value [43,44]. Arabi and his co-workers have prepared dummy molecularly-imprinted silica nanoparticles (DMISNPs) with high selectivity for AA based on the techniques of sol-gel, one-step synthesis and central composite design [45]. In the polymerization process, 3-aminopropyltrimethoxysilane (APTMS) was used as the functional monomer, propionamide as the dummy template, and tetraethyl orthosilicate (TEOS) as the crosslinking agent. The obtained DMISNPs were further used as sorbent to extract AA from food samples using a matrix solid-phase dispersion method (MSPD), and then combined with HPLC-MS to detect AA in biscuits and bread samples. The results showed that DMISNPs have high porosity, good uniformity and high selectivity and affinity for AA. More importantly, this molecularly-imprinted polymer (MIP) composite was easy to completely remove the dummy templates to obtain highly-recognized cavities, which are beneficial to eliminate template problems and improve mass transfer and extraction efficiency (Figure 1A). The MSPD method also greatly reduces the consumption of toxic organic solvents. Magnetic solid-phase extraction (MSPE) consumes less organic solvent and has higher contact-surface efficiency and repeatability. In addition, the magnetic adsorbent does not require the processes of filtration, centrifugation, and precipitation, and can be directly collected magnetically, which greatly simplifies the pretreatment steps and has received great attention in complex sample pretreatment techniques in recent years [46–48]. Nodeh successfully developed a hybrid of magnetite (Fe_3O_4) and sol-gel of TEOS and methyltrimethoxysilane (MTMOS) to modify the graphene. The obtained material was further applied as magnetic solid purified adsorbent for rapid purification and extraction of AA in various foods, combined with GC-MS analysis (Figure 1B) [49]. Compared with previous studies based on MSPE, this study used the matrix-matching method for calibration, which has better linearity (R^2 = 0.9990), lower LOD (0.061–2.89 µg kg^{-1}), and higher recovery (82.7–105.2%). The prepared Fe_3O_4@graphene-TEOS -MTMOS extractant can be reused at least seven times with a recovery rate higher than 85%. Bagheri et al. also used propionamide as a dummy template to fix a thin layer of chitosan-imprinting network on Fe_3O_4@PEG core in aqueous medium, and obtained a dummy MIP (DMIP) (Figure 1C), which was applied to detect AA in biscuit samples in combination with HPLC [50]. This DMIP had a uniform nano-core-shell structure and good magnetic properties, which were conducive to simple and rapid separation. This novel core-shell recognition material further overcame the shortcomings of poor selectivity of MSPE, and the synthesis was simple, easy to separate, in line with the green synthesis strategy, and very suitable for the pretreatment and purification of complex samples.

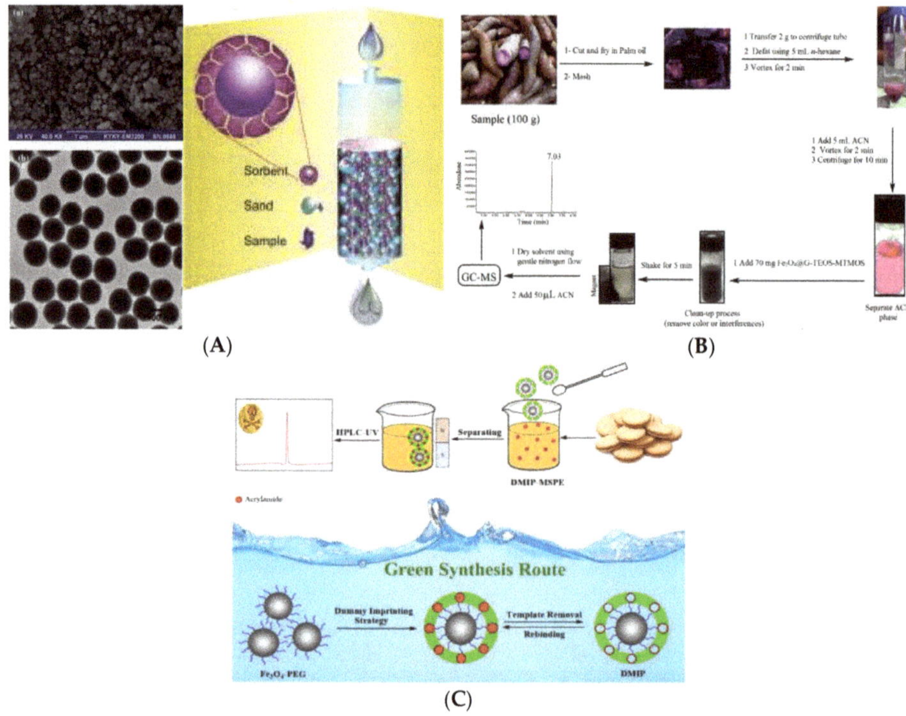

Figure 1. (**A**) Surface topography of dummy molecularly-imprinted silica nanoparticles (DMISNPs) (a: SEM and b: TEM) and the DMISNPs-matrix solid-phase dispersion method (MSPD) extraction procedure [45]. Copyright: Food Chemistry, 2016. (**B**) Schematic procedure of magnetic solid-phase extraction (MSPE) using Fe$_3$O$_4$@graphene-TEOS-MTMOS [49]. Copyright Food Chemistry, 2018; (**C**) Schematic procedure of MSPE using dummy molecularly-imprinted polymer (DMIP) with Fe$_3$O$_4$@PEG as core [50]. Copyright: Talanta, 2019.

On the other hand, solid-phase microextraction (SPME) is a kind of non-solvent selective extraction method, which abandons the shortcomings of the traditional SPE process that needs column packing and solvent for desorption, and only needs a simple syringe to complete the whole pretreatment and injection processes. Therefore, SPME has the characteristics of low cost, simple device and operation, fast, efficient, and high sensitivity. As a unique sample pretreatment and enrichment method, SPME has also been paid attention to in the detection of AA [51,52]. A direct, fast strategy based on headspace SPME has been developed for AA extraction from coffee beans [53]. The commercial SPME fiber-coated polydimethylsiloxane (PDMS) was employed to carry out the silylation reaction of AA with *N*,*O*-bis(trimethylsilyl) trifluoroacetamide and further quantified AA analysis in combination with GC-MS methods. The LOQ of AA for this method is 3 µg kg^{-1} with good reproducibility (RSD: 2.6%), which was in accordance with the EU's recommendations for monitoring AA content in foods [54]. The liquid-phase microextraction (LPME) method realizes the integration of sampling, separation, purification, concentration, and injection, which is simple and fast in AA detection [55,56]. Elahi et al. have developed a dispersive liquid microextraction combined with GC-MS method to detect AA in cookie samples [57]. This study has effectively removed the complex matrix components in sample pretreatment and significantly extracted trace amounts of target analytes in a short time. Lower values of LOD (0.6 µg kg^{-1}) and LOQ (1.9 µg kg^{-1}) and acceptable recovery range (89–95%) with RSD of 9.2% demonstrated the merits of the method in the detection of AA at low and high content in biscuits.

Stable isotope tracing technology is one technique that uses the enriched stable isotope-labeled compounds as tracers and analyzes isotopic compositions to monitor or detect certain biochemical processes [58,59]. At present, the main internal standard compounds used in the detection of AA by MS include d3-AA, $^{13}C_3$-AA, *N*, *N*-dimethylacrylamide, propionic acid, and methacrylamide [60]. By adding $^{13}C_3$-AA internal standard solution to the test sample, through a series of extraction, purification, and derivatization of bromine reagents, the GC-MS method can reach an LOD of 10 µg kg^{-1} of AA in rice [61]. Lim et al. employed the deuterated d3-AA as an internal standard for the analysis of AA content in food samples, and the established LC-MS/MS method achieved a lower LOD (0.04 µg kg^{-1}) and LOQ (0.14 µg kg^{-1}) [62]. The RSD values in the AA concentration range of 20–100 µg kg^{-1} was less than 8%, demonstrating good sensitivity and reproducibility of the developed method. This strategy did not require further extraction and purification processes, but still required a certain amount of toxic organic reagents. Ferrer-Aguirre et al. employed deuterated d5-AA as an internal standard, in combination with HPLC coupled to triple quadrupole-tandem MS, to initially determine AA content in different starchy foods (such as potato chips and potatoes) [63]. This effective analysis strategy used the water as an extraction solvent, which minimized the detection cost and reduced the sample processing. The values of LOD and LOQ were 4 and 12 µg kg^{-1} (potato chips) and 2 and 5 µg kg^{-1} (roasted asparagus), respectively. This method has the advantages of simple process, low cost, and no toxicity, and is suitable for preliminary identification of AA in different starchy foods. Carbon-labeled internal standards were also used for the detection of AA content in foods. Yoshioka Toshiaki et al. developed a supercritical fluid chromatography tandem mass spectrometry (SFC-MS/MS) technique using $^{13}C_3$-AA as an internal standard for rapid quantitative analysis of AA in various beverage, cereal, and confectionery samples [64]. Compared with methods using hydrogen-labeled internal standards, this proposed method has extremely high accuracy and sensitivity, simplifies the detection steps, and can quickly quantify low-concentration analytes, which has a very important practical value.

3. New Strategies for AA Analysis

Food belongs to fast consumer goods, which require fast detection speeds and high throughput, which puts forward new requirements for food analysis and detection. Although the traditional instrumental analysis of AA in foods has obvious advantages in detection stability and accuracy, it needs a relatively cumbersome sample pretreatment process, which makes it far behind in real-time, online and large-number sample analysis. With the rise and in-depth development of technologies such as immunity, sensing, and chips, some simple, fast, low cost, and convenient analytical strategies have been proposed and applied to the detection of AA content in foods.

3.1. Capillary Electrophoresis

Capillary electrophoresis (CE) has the characteristics of fast analysis speed and high separation efficiency, and requires a small amount of sample, making it an effective tool for the analysis of trace components in foods [65,66]. CE is based on different charge ratios of the target substance to achieve efficient separation. Therefore, the target substance is required to have a certain charge (positive or negative). The non-charged AA can achieve the detection purpose by adding an ionic surfactant to the detection system to form a charged micelle on its surface. Abd El-Hady et al. developed an analyte focusing by ionic liquid micelle collapse (AFILMC) capillary electrophoresis method combined with ionic liquid ultrasonic-assisted extraction to simultaneously measure AA, asparagine, and glucose in foods [67]. In this process, 1-butyl-3-methylimidazolium bromide (BMIM$^+$ Br$^-$) was used as a surfactant, and the washing procedure of HCl and water was appropriately optimized to sufficiently reduce the adsorption of BMIM$^+$ Br$^-$. The separation and extraction efficiency exceeded 97.0%. The AFILMC measurements achieved adequate reproducibility and accuracy with RSD 1.14–3.42% ($n = 15$) and recovery 98.0–110.0% within the concentration range of 0.05–10.0 µmol L^{-1}. The LODs achieved to 0.71 µg kg^{-1} for AA, 1.06 µg kg^{-1} for asparagine, and 27.02 µg kg^{-1} for glucose, respectively, with linearity ranged between 2.2 and 1800 μg kg^{-1}. This method has the characteristics of environmental protection, low cost, high efficiency, and high selectivity. Pre-column derivatization is another method used in CE to charge AA.

Yang et al. proposed an efficient method for AA derivatization based on thiol-olefin reaction using cysteine as a derivatization reagent, and combined with capacitively-coupled contactless conductivity detection (C^{4}D) for CE analysis of AA (Figure 2A) [68]. This method can analyze labeled AA within 2.0 min, and the RSD of migration time and peak area are less than 0.84% and 5.6%, showing good accuracy and selectivity. At the same time, the C^{4}D signal of the AA derivative has a good linear relationship with the AA concentration in the range of 7–200 μmol L^{-1} (R^2 = 0.9991), LOD and LOQ (0.16 μmol L^{-1} and 0.52 μmol L^{-1}). Due to the advantages of simple sample pretreatment, high derivatization efficiency, short analysis time, and high selectivity and sensitivity, this CE-C^{4}D is expected to achieve further miniaturization for field analysis.

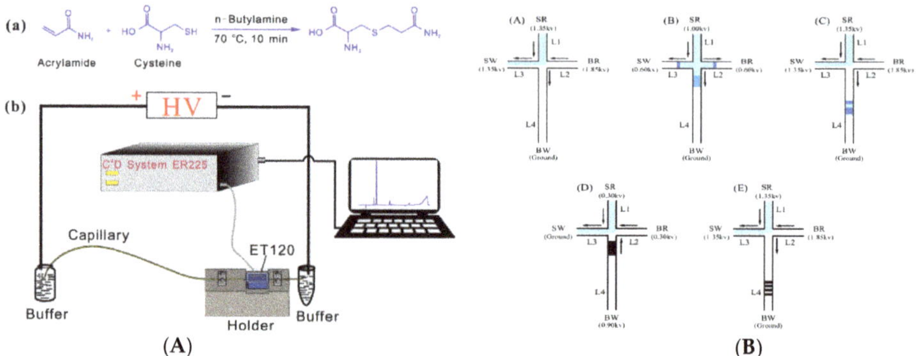

Figure 2. (**A**) Schematic illustration for thiol-ene click derivatization of acrylamide (AA) using cysteine and the CE-C^{4}D system [68]. Copyright: Journal of Agricultural and Food Chemistry, 2019. (**B**) Five-steps of microchip electrophoresis technology (MCE) strategy. A: preloading, B: loading, C: prolonged field-amplified sample stacking, D: reversed-field stacking, and E: separation [69]. Copyright: Food Chemistry, 2016.

A portable microchip requires a small amount of detection samples, especially when combined with electrophoresis technology, which shortens the separation channel, thus achieving faster separation and more sensitive detection [70,71]. Because the content of AA in foods is very low, it is not suitable for microchip electrophoresis technology (MCE). It must be combined with on-line enrichment technology to improve the sensitivity. This on-line enrichment and detection method effectively overcomes the interference of food complex matrix and improves the detection speed [72,73]. Wu et al. proposed an MCE based on a combination of high-field amplification and anti-field superposition of online multiple pre-enrichment technology for efficient analysis of AA in foods (Figure 2B) [69]. The best separation has been achieved under the condition of 100 mmol L^{-1} borate solution at pH 9.3 as the running buffer. The sensitivity of this method (LOD: 1 μg L^{-1}) is 41–700 times higher than the previously-reported CE of on-line preconcentration technology, which has been successfully applied to detect AA content in potato chips and French fries with reliable results and satisfactory recoveries. Compared with traditional methods for AA detection, this effective method has the advantages of short analysis time, low sample and reagent consumption, and low instrumental cost.

3.2. Immunoassay Method

Immunoassay is one new, rapid, and high-throughput analysis strategy based on the specific combination of antigen (Ag) and antibody (Ab). After nearly 20 years of development, the immunoassay has gradually developed into enzyme-linked immunoassay (ELISA) [74,75], chemiluminescent immunoassay [76,77], fluorescent immunoassay [78,79] and so on, which have been widely used in food analysis, especially in the field of rapid detection. The Ab with specific binding ability is the basis of immunoassay. Due to the fact that AA is a small molecular compound, lacking in antigenic determinant and immunogenicity, AA is usually cross-linked with the carrier proteins, bovine serum albumin (BSA), ovalbumin (OVA), etc., with immune response, to prepare incomplete antigens, and the polyclonal Abs with specific recognition ability are further obtained by immunizing animals. Singh et al. prepared polyclonal Abs which were raised against a hapten derived from AA and 3-mercaptobenzoic acid (3-MBA) and established an indirect competitive ELISA (ic-ELISA) to quickly quantify AA in complex foods matrix and water [80]. This ic-ELISA had high affinity and specificity for AA-3-MBA derivatives and did not cross-react with the main precursors (asparagine, aspartic acid, AA, or 3-MBA) that form AA in foods. The LODs achieved for AA-3-MBA in food matrices and water were 5.0 μg kg^{-1}, and 0.1 μg L^{-1}, respectively, which verified that the developed ic-ELISA has extremely high sensitivity and good AA recovery, and is suitable for AA detection in multiple matrices. Wu and his co-workers used the 4-mercaptophenylacetic-acid-derived AA (AA-4-MPA) to prepare polyclonal Abs and developed a pre-analytical derivatization method for ic-ELISA analysis of AA (Figure 3A) [81]. By comparison with the results from the HPLC-MS/MS method, this ic-ELISA has better accuracy and reliability (IC$_{50}$: 2.86 μg kg^{-1}, LOD: 0.036 μg kg^{-1}, linear range: 0.25–24.15 μg kg^{-1}), lower detection cost, and is suitable for routine rapid screening of AA in food samples. Monoclonal antibodies (MAbs) are homologous Abs produced by a B-cell clone that recognize an antigenic determinant, have high titer, strong homogeneity and specificity, and low cross reactivity. Zhu et al. used the 4-mercaptobenzoic acid-derived AA (AA-4-MBA) to couple to carrier proteins (BSA and OVA) (Figure 3B) [82]. The resulting conjugates of AA-4-MBA–BSA and AA-4-MBA–OVA were used as the immunogen and coating antigen. The obtained MAb is not specific for AA or 4-MBA but has high affinity for AA-4-MBA (IC$_{50}$: 32 μg L^{-1}; LOD 8.87 μg L^{-1}). The quantitative working range is 8.87–12.92 μg L^{-1} (IC$_{20}$ to IC$_{80}$) and the cross-reactivity with other analogs is less than 10%, meaning that the developed ic-ELISA method has extremely high specificity and can effectively detect AA in high-temperature-cooking foods.

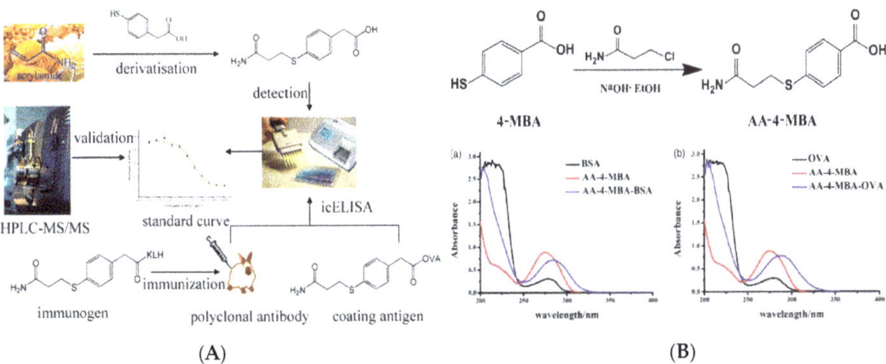

Figure 3. (**A**) The indirect competitive ELISA (ic-ELISA) procedure for AA analysis in foods [81]. Copyright: Journal of Agricultural and Food Chemistry, 2014. (**B**) Scheme for 4-mercaptobenzoic acid-derived AA (AA-4-MBA) hapten synthesis and UV scans of curves of AA-4-MBA, protein, and Ag [82]. Copyright: Food and Agricultural Immunology, 2016.

Compared with the ELISA, the immunochromatographic strip (ICS) is a relatively mature integrated detection product, which is simpler to operate and can be completed without professional operators, meeting the real-time and fast-detection requirements of AA. Assaat et al. have produced, purified, and characterized a polyclonal Ab against AA for ICS testing of AA [83]. Polyclonal anti-AA Ab was prepared by injecting *N*-acryloxysuccinimide conjugated BSA hapten into New Zealand white rabbits and further purified with protein A and conjugated with Au nanoparticles (AuNPs). According to the obtained results, the ICS prepared in this study quantitatively showed that the intensity of the red line increased with the increase of AA concentration, and was sensitive to standard AA solution at 1 g L^{-1} concentration, which is expected to be applied for the rapid detection of AA in foods.

Immunoassay based on biological Abs has the advantage of fast detection speed and high throughput. However, the process of obtaining biological Abs with high specific binding ability is complicated and costly; it is easy to be affected by environmental conditions in the detection process, resulting in false positive results, which to some extent limit the development of biological immunoassay. Molecular imprinting technology (MIT) can chemically synthesize high-specific and stable polymers based on the principle of Ab formation. The obtained MIPs also called "artificial antibodies", are used in the development of biomimetic ELISA method, which has very broad application prospects. A direct competitive biomimetic ELISA rapid analysis method for AA analysis was developed by Sun et al. using a hydrophilic-imprinted membrane as a biomimetic Ab [84]. In the preparing process of the imprinted membrane, -COOH of methacrylic acid reacted with -NH$_2$ of AA, and an imprinted cavity and a specific binding site of -OH group were generated in a predetermined direction, so that the imprinted membrane had high binding and selectivity to AA. The developed biomimetic ELISA method had high sensitivity (IC$_{50}$: 8.0 ± 0.4 mg L^{-1}) and low LOD (IC$_{15}$: 85.0 ± 4.2 µg L^{-1}), and for an AA-blank potato sample, the recovery rate ranged from 90.0% to 111.5%. The biomimetic ELISA method is simple in pretreatment and does not require Ab coating and BSA/PBS blocking procedures, which greatly reduces the operation time. Additionally, the developed blotting membrane can be reused 20 times without loss of sensitivity, which greatly reduces the cost of analysis.

3.3. Sensor Analysis Technique

Sensors can on-line monitor the binding reaction between the tested substance and the recognition element, and convert the generated binding signal into a signal that can be processed quantitatively, such as electricity, light, or mass, to achieve the purpose of analysis and detection [85,86]. The AA molecule contains a -NH$_2$ structure, which can be hydrolyzed to NH$_4$$^+$ and then detected by the selective electrode [87]. Because this method is based on the catalytic hydrolysis of -NH$_2$, it has a strong cross-reaction to compounds containing -NH$_2$ groups such as formamide and acetamide [88]. A new two-step waveform containing a process of separation of reverse-phase LC coupled to a pulsed amperometric detection was reported by Casella's group for the quantification of low concentrations of AA in foodstuffs such as coffee and potato fries. Compared to the classical type of waveform, the proposed two-step waveform showed favorable analytical performance in terms of LOD (1.4 µg kg^{-1}), precision, and improved long-term reproducibility.

3.3.1. Electrochemical Sensing Analysis Based on Biomolecules

As an emerging analysis strategy, biosensors have made in-depth developments in the fields of environment, medicine, and food. In food safety, various types of biosensors are designed for the analysis of food components and harmful substances [89–92]. Hemoglobin (Hb) is a redox-active protein that involves four polypeptide chains, each of which has an electroactive group of Fe^{3+}/heme. The electrical activity of Hb is related to the reversible conversion of Hb-Fe^{3+} to Hb-Fe^{2+} [93,94]. At the same time, the valine α-NH$_2$ in Hb can be combined with AA to form a complex, which causes the amount of Hb-Fe^{2+} to decrease, resulting in the change in the electron transfer on the surface of the sensing electrode. Based on this principle, the Hb can be modified on the surface of the transducer for AA detection [95]. Compared with other proteins with similar mechanisms of action (myoglobin, cytochrome c, etc.), Hb is more appropriate in the construction of a AA biosensor, due to its commercial accessibility at low cost, its relatively higher stability, and its configuration (*N*-(2-carbamoyl-ethyl)-*L*-valine), which is similar to that of the glycidamide (*N*-(2-carbamoyl-2-hydroxyethyl) -*RS*-valine), which facilitates the formation of Hb–AA adduct [96,97]. However, Hb has a complex spatial structure, and its electroactive center exists inside a polypeptide chain, which easily causes the electrode surface to be passivated and slows down the electron transfer rate. In recent years, in view of these shortcomings and problems, some meaningful solutions have been proposed.

Yadav et al. prepared one kind of Hb nanoparticle (HbNP) by the desolvation method, and covalently immoblized HbNPs on a polycrystalline Au electrode to construct a current-type AA biosensor (Figure 4) [98]. At the experimental conditions of 20 °C, 0.26 V, and pH 5.0, the HbNP's modified Au electrode showed the best current response within 2 s. In the water extract of foods at spiked AA concentration of 5 and 10 mmol L^{-1}, a remarkable recovery of more than 95% was achieved, and the intra- and inter-assay coefficients of variation were lower than 5%. The wide-working range (0.1–100 nmol L^{-1}) and the lower LOD (0.1 nmol L^{-1}) signified this HbNP's modified Au electrode could offer an effective measurement of AA in various processed foods. In this research, the use of HbNPs instead of natural Hb molecules solved the problem that Hb easily causes the electron transfer rate on the electrode surface to slow down, increases the specific surface area, and enables highly-sensitive micro detection of AA. In addition, the constructed sensor was not affected by the structural analogs of AA (such as acrylic acid and propionic acid) during the working process, signifying its good selective recognition ability. Asnaashari and his co-workers designed an effective double-stranded DNA (dsDNA)/Hb-modified screen-printed Au electrode for the detection of AA (Figure 5A) [99]. The square wave voltammetry (SWV) was used to monitor the current response of the designed biosensor to AA and AA-valine adduct, as well as the changes in the reduction/oxidation process of Hb-Fe^{3+}/Hb-Fe^{2+}. This fabricated sensor obtained a linear working range of AA (2.0 × 10^{-6} to 5.0 × 10^{-2} mol L^{-1}) and a lower LOD (1.58 × 10^{-7} mol L^{-1}) and had good reproducibility and high stability, which is suitable for direct determination of AA in foods.

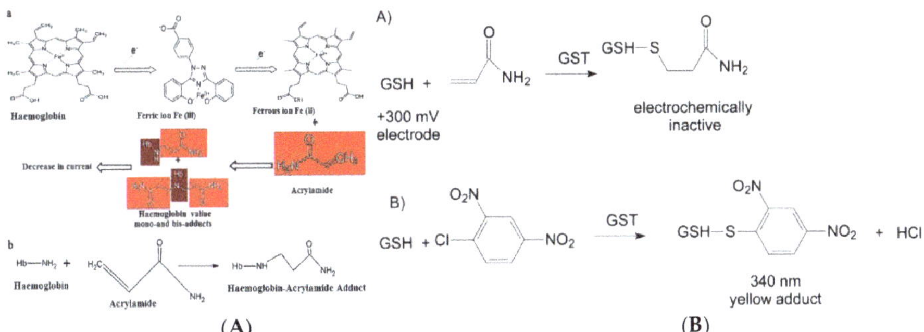

Figure 4. (**A**) a: Electrochemical reactions involved in the functioning of the hemoglobin nanoparticle (HbNP)-AA biosensor and b: adduct formation of HbNPs and AA. (**B**) Schematic representation of chemical reaction of the fabrication of HbNPs onto an Au electrode [98]. Copyright: International Journal of Biological Macromlecules, 2018.

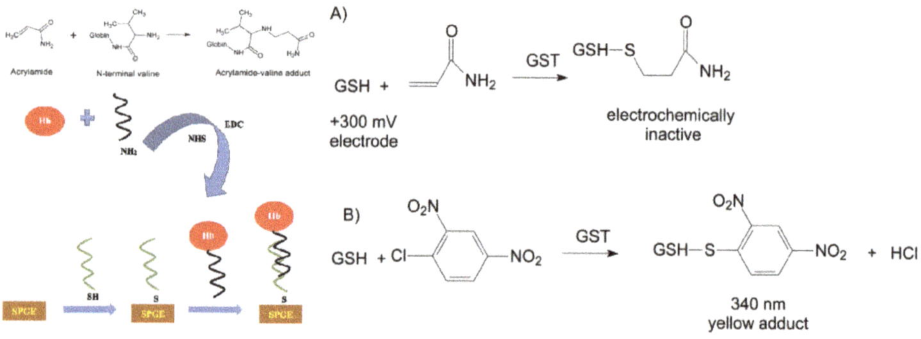

Figure 5. (A) Hb–AA adduct formation and schematic description of the preparation of dsDNA/Hb-modified screen-printed Au electrode for AA [99]. Copyright: Food Chemistry, 2019. (B) The reactions catalyzed by glutathione *S*-transferase (GST). A: coupling of GSH with AA and B: color reaction used for enzymatic activity measurements [100]. Copyright: Rsc Advances, 2018.

Studies have shown that most of the AA ingested in the body is metabolized in the liver, except for a small part (<10%) which is excreted in the prototype with urine [101–104]. Under the action of enzymes, AA can combine with glutathione (GSH) to form thioglycolic acid compound, which is further converted into glycidomide (GA). In the case of low AA dose, about 50% of AA will be converted to GA, while in the case of high dose AA, most of AA will react with GSH and about 13% will be converted to GA. Therefore, GSH can also be modified on the electrode surface for the sensing response of AA in foods. Bucur et al. have proposed a method based on the amperometric monitoring of the coupling reaction between reduced glutathione (GSH) and AA catalyzed by glutathione *S*-transferase (GST) to produce an electrochemically-inactive compound (Figure 5B) [100]. Cobalt phthalocyanine was modified on a screen-printed electrode to monitor the decrease in GSH concentration at +300 mV, further aimed to detect the target AA. At the optimal GSH concentration (100 μmol L^{-1}), the linear range for AA analysis was 7–50 μmol L^{-1} and LOD achieved to 5 μmol L^{-1}. This proposed method was simple, did not require auxiliary substrates such as *1-chloro-2,4-dinitrobenzene* (CDNB), and did not need to suppress adverse competitive kinetics. The whole detection process was not affected by interfering compounds usually found in foods and could be applied for real sample analysis.

Experimental factors such as the electron transfer rate on the electrochemical sensing interface, and the immobilization capacity and biological activity of the identification element have profoundly affected the performance of the sensors. The introduction of nanomaterials not only increases the electron transfer rate, but also increases the fixed amount and activity of biometric recognition elements, which can significantly improve the stability and sensitivity of sensors. Wulandar et al. developed a platinum (Pt) and Hb-modified boron-doped diamond electrode (Pt-BDD) for the construction of AA biosensors (Figure 6A) [105]. The surface of Pt-BDD modified with PtNPs had excellent stability. Meanwhile, Hb-Pt-modified BDD (Hb-Pt-BDD) showed a linear CV response in acetate buffered saline (0.2 mol L^{-1}, pH 4.8) with AA concentration range of 0.01–1 nmol L^{-1}. The LOD and LOQ achieved to 0.0085 nmol L^{-1} and 0.026 nmol L^{-1}, respectively. These results demonstrated that the prepared Hb-Pt-BDD electrode has high stability, good sensitivity, and is reusable because it removes Hb adducts without removing Pt on the surface of BDD. Compared with PtNPs, Au nanomaterials also have excellent catalytic activity, efficient electron transfer performance, and good optical characteristics, and have been widely used in the sensing fields. Figure 6B shows the development of an ultrasensitive immunosensor using chitosan/SnO$_2$-SiC hollow-sphere nanochains/AuNPs as signal amplification for detecting AA in water and food samples [106]. SnO$_2$-SiC hollow-sphere nanochains with high surface area and AuNPs with good electrical conductivity were prepared on the surface of glassy carbon electrodes pre-coated with chitosan for subsequent fixed coating of antigens. Under this working mode, the constructed immunosensor has a lower LOD of 45.9 ± 2.7 ng kg^{-1} and wider working range of 187 ± 12.3 ng kg^{-1} to 104 ± 8.2 mg kg^{-1} at the optimized conditions. The recovery of AA in the spiked samples was in the range of 86.0–115.0%. The immunosensor exhibited a sensitive response to AA, and acceptable repeatability and stability.

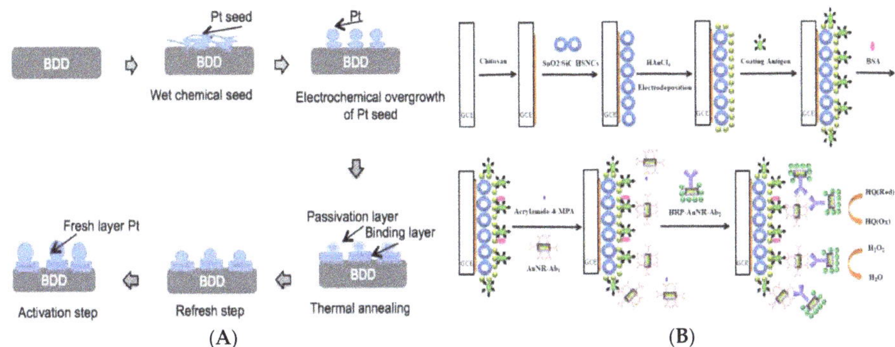

Figure 6. (**A**) Schematic representation of the Pt modification of a boron-doped diamond electrode (BDD) [105]. Copyright: Sensors and Materials, 2019. (**B**) Schematic illustration of the AA immunosensor fabrication [106]. Copyright: Analytica Chimica Acta, 2019.

Carbon-based nanomaterials (carbon nanotubes, nanocarbon spheres, graphene, carbon nanofibers, etc.) have unique conductive and good mechanical properties, such as unique nanometer size, high surface area, high electron transfer rate, high stability and the ability to be modified on the surface that can maintain the stability and activity of biorecognition molecules [107,108]. Liu et al. have combined the composite of AuNPs-multi-walled carbon nanotubes (MWCNTs)-chitosan (AuNPs-MWCNTs-CS) with sol-gel MIT to construct a molecularly-imprinted electrochemical sensor for AA detection [109]. The composite of AuNPs and MWCNTs was introduced to improve the polymer conductivity and expand the surface area of electrode. At the working potential of 0–0.4 V, this developed electrochemical sensor exhibits a linear current response to AA in the concentration range of 0.05–5 mg L^{-1} with LOD of 0.028 mg L^{-1} (S/N = 3). With the characteristics of good repeatability, stable and reliable storage, good selectivity, high sensitivity, and low cost, this molecularly-imprinted electrochemical sensor has a very broad application prospect.

Carbon nanomaterials and their composites are also used in Hb-based AA electrochemical biosensors to enhance the surface electron transfer rate of electrodes. Batra et al. have developed an Hb electrochemical biosensor based on the composite of carboxylated MWCNTs/CuNPs/polyaniline (PANI) (c-MWCNTs/CuNPs/PANI), which can detect AA with high sensitivity (Figure 7A) [110]. Under the optimized experimental conditions, the fabricated sensor has low LOD (0.2 nmol L^{-1}), high sensitivity (72.5 µA nmol L^{-1} cm^{-2}), fast response time (<2 s), and wide linear range (5 nmol L^{-1} to 75 mmol L^{-1}). When stored at 4 °C, the electrode can be used 120 times in 100 days with acceptable repeatability and stability. Varmiraa et al. also constructed an effective electrochemical biosensor based on Hb-dimethyldioctadecyl ammonium bromide (DDAB)/Pt-Au-palladium three metallic alloy NPs chitosan-1-/ethyl-3-methylimidazolium bis(trifluoromethylsulfonyl)imide/MWCNTs-IL/glassy carbon electrode (Hb-DDAB/PtAuPd NPs/Ch-IL/MWCNTs-IL/GCE) for selective and sensitive determination of AA in food samples (Figure 7B) [111]. The developed sensor can determine AA in two linear concentration ranges of 0.03–39.0 nmol L^{-1} and 39.0–150.0 nmol L^{-1} using SWV with LOD of 0.01 nmol L^{-1} and can selectively detect the target AA even in the presence of high concentrations of common interferences, confirming its highly selectivity. From the experimental results, it has been confirmed that the proposed sensor has a short response time (<8 s), good sensitivity, long-term stability, repeatability and reproducibility, and is capable of successfully measured AA in potato chips.

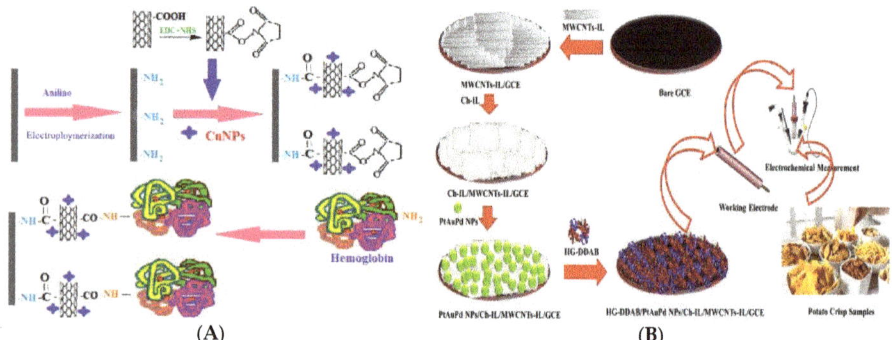

Figure 7. (**A**) Schematic representation of the fabrication of Hb/c-MWCNTs/CuNPs/PANI/PG [110]. Copyright: Analytical Biochemistry, 2013. (**B**) Schematic representation of the biosensor based on Hb-DDAB/PtAuPd NPs/Ch-IL/MWCNTs-IL/GCE [111]. Copyright: Talanta, 2018.

In summary, electrochemical sensing methods for AA detection have certain selectivity, high sensitivity, good reproducibility, and cannot be interfered with by other ingredients in foods. This gives them a unique advantage in the analysis of AA content in foods.

3.3.2. Fluorescence Sensing Analysis Method

Fluorescence sensors express signals generated by molecular recognition in the form of fluorescence (changes in fluorescence intensity and wavelength) to achieve information transmission [112,113]. Due to the merits of high sensitivity, good selectivity, and convenient use, fluorescence sensing analysis has been widely used and has made great progress in recent years [114–116]. Quantum dots (QDs) have unique photophysical properties such as high fluorescence quantum yield, size-controlled fluorescence, and photobleaching resistance, and have been widely used in the field of fluorescence sensing [117,118]. Since the AA molecule does not have fluorescent properties, it needs to be detected using other fluorescent substances. This kind of chemical reaction-based fluorescence sensing detection mode needs further research in terms of enhancing the detection sensitivity and selectivity. Especially when introducing new nanomaterials with unique properties into the fluorescence detection system, it is possible to develop more efficient and accurate analysis strategies [119–121]. Hu et al. have proposed a fluorescence sensing method for online detection of AA in potato chips, which was based on the increase of the distance between QDs caused by AA polymerization (Figure 8A) [122]. The UV light irradiation caused the C=C bond polymerization of *N*-acryloxysuccinimide-modified QDs, which shortened the distance between the QDs, leading to a decrease in fluorescence intensity. When AA is present in the tested sample, AA would participate in the above polymerization reaction, causing an increase of fluorescence intensity. The linear range and LOD of the established method reached 3.5×10^{-5}–3.5 g L^{-1} (R^2 = 0.94) and 3.5×10^{-5} g L^{-1}, respectively. Although the sensitivity and specificity of this method cannot be compared with those of standard instrumental analysis, it greatly reduces the cost and time of detection and is suitable for the rapid detection of AA online in food processing. The ZnS QDs doped with Mn^{2+} were added onto graphene oxide as a fluorescent source to prepare an AA-MIP, which was successfully used as an environmentally-friendly fluorescent probe for AA detection [123]. When AA was adsorbed by AA-MIP, the fluorescence of ZnS QDs-doped was quenched, and the quenching effect was much stronger than that of non-imprinted polymers. The excitation and emission spectra of AA-MIP peaked at 325 nm and 601 nm, respectively. Under the optimal experimental conditions, the fluorescence of ZnS QDs in AA-MIP decreased within a linear range of 0.5–60 µmol L^{-1} AA concentration, and the LOD reached 0.17 µmol L^{-1}. In the spiked water, acceptable recovery in a range of 100.2–104.5% with remarkable RSD of 1.9–3.9% was obtained, signifying the great accuracy and sensitivity of the developed methods.

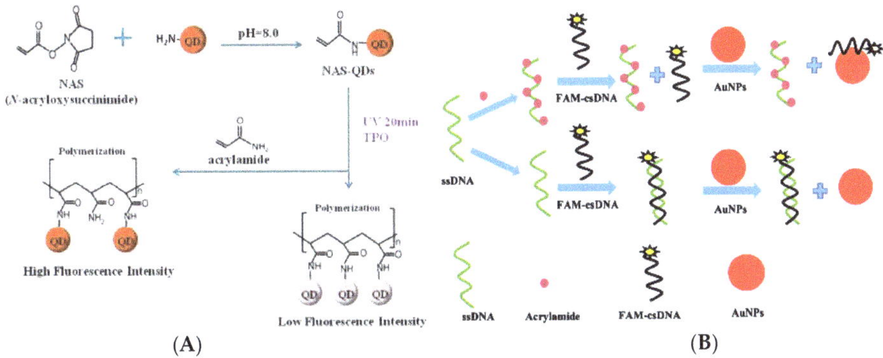

Figure 8. Schematic representation of fluorescent methods for AA detection. (**A**) CdSe/ZnS quantum dots (QDs) [122]. Copyright: Biosensors & Bioelectronics, 2014. (**B**) AuNPs and FAM-dsDNA [124]. Copyright: Sensors and Actuators B-Chemical, 2018.

Because of their excellent optical and catalytic properties, easy synthesis, high chemical stability and selectivity, and high absorption coefficient, colloidal AuNPs are widely used as fluorescence quenchers in fluorescence sensors [125]. Figure 8B shows a simple, fast, and accurate fluorescence sensor using AuNPs and FAM-labeled dsDNA (FAM-dsDNA) for AA detection [124]. The detection principle was that the AA target present in the environment forms an adduct with single-stranded DNA, and the freely-existing FAM-labeled complementary strand DNA was adsorbed on the surface of AuNPs, causing the AuNPs quench. This proposed fluorescence sensor could be quickly assembled and complete the detection process in a short time, and have high sensitivity and selectivity, and wide linear response range of AA (1×10^{-7}–0.05 mol L^{-1}) and LOD of 1×10^{-8} mol L^{-1}.

4. Conclusions

In recent years, various strategies based on different principles have been successfully developed for the analysis of AA content in different food matrices. In contrast, the traditional instrumental analysis strategies based on chromatographic separation and mass spectrometry are still the first choice for AA analysis, due to the advantages of high accuracy and good reproducibility. In follow-up research, the development of efficient, stable, cheap, and convenient pretreatment purification materials for food matrices will continue to be one of the research hotspots. ELISA analysis kits and immunoassay test strips with the characteristics of high throughput and low cost have broad application prospects in rapid screening of large numbers of samples, but they need to be improved in terms of detection stability and environmental adaptability. Electrochemical and fluorescence sensing technologies need further research in the construction of sensing interfaces and the improvement of stability. The excellent electrical and optical properties of various nanomaterials provide new ideas for developing nano-sensing methods with high sensitivity, high throughput, and good reproducibility.

Author Contributions: M.P. provided the idea and financial support of the research and completed Section 4; K.L. coordinated and organized the writing of the entire manuscript and completed the Section 1, Section 2, Section 3.1, and part of Section 3.3 (M.P. and X.L. contributed to this article equally); J.Y. completed Section 3.2. and checked the language and format of the manuscript; L.H. and X.X. completed part of Section 3.3 and analyzed and compared the results of the experiment; S.W. provided the framework of the paper and finally checked the quality of the article. All authors have read and agreed to the published version of the manuscript.

Funding: This research was funded by the National Key R&D Program of China (No. 2017YFC1600402), National Natural Science Foundation of China (No. 31972147), Tianjin Technical Expert Project (No. 19JCTPJC52700),

Foods **2020**, *9*, 524

Project of Tianjin Science and Technology Plan (No. 18ZYPTJC00020), and the Open Project Program of State Key Laboratory of Food Nutrition and Safety, Tianjin University of Science and Technology (SKLFNS-KF-201907). The APC was funded by Project of Tianjin Science and Technology Plan (No. 18ZYPTJC00020).

Conflicts of Interest: The authors declare no conflict of interest.

References

1. Friedman, M. Chemistry, biochemistry, and safety of acrylamide. A review. *J. Agric. Food Chem.* **2003**, *51*, 4504–4526. [CrossRef] [PubMed]
2. Altunay, N.; Gürkan, R.; Orhan, U. A preconcentration method for indirect determination of acrylamide from chips, crackers and cereal-based baby foods using flame atomic absorption spectrometry. *Talanta* **2016**, *161*, 143–150. [CrossRef]
3. IARC. *IARC Monographs on the Evaluation of Carcinogenic Risks to Humans*; IARC: Lyon, France, 1994; Volume 60, pp. 389–433.
4. Tareke, E.; Rydberg, P.; Karlsson, P.; Eriksson, S.; Törnqvist, M. Analysis of acrylamide, a carcinogen formed in heated foodstuffs. *J. Agric. Food Chem.* **2002**, *50*, 4998–5006. [CrossRef] [PubMed]
5. Mottram, D.S.; Wedzicha, B.L.; Dodson, A.T. Acrylamide is formed in the maillard reaction. *Nature* **2002**, *419*, 448–449. [CrossRef] [PubMed]
6. Stadler, R.H.; Varga, N.; Robert, F.; Hau, J.; Guy, P.A.; Robert, M.C.; Riediker, S. Acrylamide from Maillard reaction products. *Nature* **2002**, *419*, 449–450. [CrossRef]
7. Claeys, W.L.; De Vleeschouwer, K.; Hendrickx, M.E. Kinetics of acrylamide formation and elimination during heating of an asparagine-sugar model system. *J. Agric. Food Chem.* **2005**, *53*, 9999–10005. [CrossRef]
8. Ehling, S.; Hengel, M.; Shibamoto, T. Formation of acrylamide from lipids. In *Chemistry and Safety of Acrylamide in Food*; Friedman, M., Mottram, D., Eds.; Advances in Experimental Medicine and Biology; Springer: Boston, MA, USA, 2005; Volume 561, pp. 223–233.
9. Perez-Nevado, F.; Cabrera-Banegil, M.; Repilado, E.; Martillanes, S.; Martin-Vertedor, D. Effect of different baking treatments on the acrylamide formation and phenolic compounds in Californian-style black olives. *Food Control* **2018**, *94*, 22–29. [CrossRef]
10. Ehling, S.; Shibamoto, T. Correlation of acrylamide generation in thermally processed model systems of asparagine and glucose with color formation, amounts of pyrazines formed, and antioxidative properties of extracts. *J. Agric. Food Chem.* **2005**, *53*, 4813–4819. [CrossRef]
11. Casado, F.J.; Montano, A. Influence of processing conditions on acrylamide content in black ripe olives. *J. Agric. Food Chem.* **2008**, *56*, 2021–2027. [CrossRef]
12. Charoenprasert, S.; Mitchell, A. Influence of California-style black ripe olive processing on the formation of acrylamide. *J. Agric. Food Chem.* **2014**, *62*, 8716–8721. [CrossRef]
13. Esposito, F.; Fasano, E.; De Vivo, A.; Velotto, S.; Sarghini, F.; Cirillo, T. Processing effects on acrylamide content in roasted coffee production. *Food Chem.* **2020**, *319*, 126550. [CrossRef]
14. Commission Recommendation (EU) 2017/2158. Commission Regulation (EU) 2017/2158 of 20 November 2017 establishing mitigation measures and benchmark levels for the reduction of the presence of acrylamide in food. *Off. J. Eur. Union* **2017**, *L304*, 24–44.
15. Oracz, J.; Nebesny, E.; Zyzelewicz, D. New trends in quantification of acrylamide in food products. *Talanta* **2011**, *86*, 23–34. [CrossRef] [PubMed]
16. Hu, Q.Q.; Xu, X.H.; Fu, Y.C.; Li, Y.B. Rapid methods for detecting acrylamide in thermally processed foods: A review. *Food Control* **2015**, *56*, 135–146. [CrossRef]
17. Commission Recommendation (EU) 2019/1888. *Commision Recommendation (EU) 2019/1888 of 7 November 2019 on the Monitoring of the Presence of Acrylamide in Certain Foods*; European Union: Brussels, Belgium, 2019; Volume L291, pp. 31–33, (document 32019H1888).
18. Gokmen, V.; Palazoglu, T.K. Acrylamide Formation in Foods during Thermal Processing with a Focus on Frying. *Food Bioprocess Technol.* **2008**, *1*, 35–42. [CrossRef]
19. El-Assouli, S.M. Acrylamide in selected foods and genotoxicity of their Extracts. *J. Egypt Public Health Assoc.* **2009**, *84*, 371–392. [PubMed]
20. Erkekoglu, P.; Baydar, T. Toxicity of acrylamide and evaluation of its exposure in baby foods. *Nutr. Res. Rev.* **2010**, *23*, 323–333. [CrossRef] [PubMed]

21. Kumar, J.; Das, S.; Teoh, S.L. Dietary acrylamide and the risks of developing cancer: Facts to ponder. *Front. Nutr.* **2018**, *5*, 14. [CrossRef]

22. Koszucka, A.; Nowak, A.; Nowak, I.; Motyl, I. Acrylamide in human diet, its metabolism, toxicity, inactivation and the associated European Union legal regulations in food industry. *Crit. Rev. Food Sci.* **2019**. [CrossRef]

23. Tardiff, R.G.; Gargas, M.L.; Kirman, C.R.; Carson, M.L.; Sweeney, L.M. Estimation of safe dietary intake levels of acrylamide for humans. *Food Chem. Toxicol.* **2010**, *48*, 658–667. [CrossRef]

24. Singh, P.; Singh, P.; Raja, R.B. Determination of acrylamide concentration in processed food products using normal phase high-performance liquid chromatography (HPLC). *Afr. J. Biotechnol.* **2010**, *9*, 8085–8091.

25. Geng, Z.M.; Wang, P.; Liu, A.M. Determination of acrylamide in starch-based foods by HPLC with pre-column ultraviolet derivatization. *J. Chromatogr. Sci.* **2011**, *49*, 818–824. [CrossRef]

26. Xu, L.H.; Zhang, L.M.; Qiao, X.G.; Xu, Z.X.; Song, J.M. Determination of trace acrylamide in potato chip and bread crust based on SPE and HPLC. *Chromatographia* **2012**, *75*, 269–274.

27. Sun, S.Y.; Fang, Y.; Xia, Y.M. A facile detection of acrylamide in starchy food by using a solid extraction-GC strategy. *Food Control* **2012**, *26*, 220–222. [CrossRef]

28. Yao, W.J. Direct determination of acrylamide in food by gas chromatography with nitrogen chemiluminescence detection. *J. Sep. Sci.* **2015**, *38*, 2272–2277.

29. Saraji, M.; Javadian, S. Single-drop microextraction combined with gas chromatography-electron capture detection for the determination of acrylamide in food samples. *Food Chem.* **2019**, *274*, 55–60. [CrossRef]

30. Zhang, Y.; Ren, Y.; Jiao, J.; Li, D.; Zhang, Y. Ultra high-performance liquid chromatography-tandem mass spectrometry for the simultaneous analysis of asparagine, sugars, and acrylamide in Maillard reactions. *Anal. Chem.* **2011**, *83*, 3297–3304. [CrossRef]

31. De Paola, E.L.; Montevecchi, G.; Masino, F.; Garbini, D.; Barbanera, M.; Antonelli, A. Determination of acrylamide in dried fruits and edible seeds using QuEChERS extraction and LC separation with MS detection. *Food Chem.* **2017**, *217*, 191–195. [CrossRef]

32. Fernández, A.; Talaverano, M.I.; Pérez-Nevado, F.; Boselli, E.; Cordeiro, A.M.; Martillanes, S.; Foligni, R.; Martín-Vertedor, D. Evaluation of phenolics and acrylamide and their bioavailability in high hydrostatic pressure treated and fried table olives. *J. Food Process Pres.* **2020**. [CrossRef]

33. Cagliero, C.; Ho, T.D.; Zhang, C.; Bicchi, C.; Anderson, J.L. Determination of acrylamide in brewed coffee and coffee powder using polymeric ionic liquid-based sorbent coatings in solid-phase microextraction coupled to gas chromatography-mass spectrometry. *J. Chromatogr. A* **2016**, *1449*, 2–7. [CrossRef] [PubMed]

34. Luo, L.; Ren, Y.; Liu, J.; Wen, X.D. Investigation of a rapid and sensitive non-aqueous reaction system for the determination of acrylamide in processed foods by gas chromatography-mass spectrometry. *Anal. Methods* **2016**, *8*, 5970–5977. [CrossRef]

35. Jozinovic, A.; Sarkanj, B.; Ackar, D.; Balentic, J.P.; Subaric, D.; Cvetkovic, T.; Ranilovic, J.; Guberac, S.; Babic, J. Simultaneous determination of acrylamide and hydroxymethylfurfural in extruded products by LC-MS/MS method. *Molecules* **2019**, *24*, 1971. [CrossRef] [PubMed]

36. Fernandes, C.L.; Carvalho, D.O.; Guido, L.F. Determination of acrylamide in biscuits by high-resolution orbitrap mass spectrometry: A novel application. *Foods* **2019**, *8*, 597. [CrossRef] [PubMed]

37. Calbiani, F.; Careri, M.; Elviri, L.; Mangia, A.; Zagnoni, I. Development and single-laboratory validation of a reversed-phase liquid chromatography-electrospray-tandem mass spectrometry method for identification and determination of acrylamide in foods. *J. AOAC Int.* **2004**, *87*, 107–115. [CrossRef] [PubMed]

38. Galuch, M.B.; Magon, T.F.S.; Silveira, R.; Nicacio, A.E.; Pizzo, J.S.; Bonafe, E.G.; Maldaner, L.; Santos, O.O.; Visentainer, J.V. Determination of acrylamide in brewed coffee by dispersive liquid-liquid microextraction (DLLME) and ultra-performance liquid chromatography tandem mass spectrometry (UPLC-MS/MS). *Food Chem.* **2019**, *282*, 120–126. [CrossRef] [PubMed]

39. Tolgyesi, A.; Sharma, V.K. Determination of acrylamide in gingerbread and other food samples by HILIC-MS/MS: A dilute-and-shoot method. *J. Chromatogr. B* **2020**, *1136*, 121933. [CrossRef] [PubMed]

40. Xiong, J.; Qian, S.; Xie, Y.H.; Xie, Z.W.; Li, H.X. Simultaneous determination of acrylamide, aniline and benzidine in water sample by high performance liquid chromatography-tandem mass spectrometry. *Chin. J. Anal. Chem.* **2014**, *42*, 93–98.

41. Lee, K.J.; Lee, G.H.; Kim, H.S.; Oh, M.S.; Chu, S.; Hwang, I.J.; Lee, J.Y.; Choi, A.; Kim, C.I.; Park, H.M. Determination of heterocyclic amines and acrylamide in agricultural products with liquid chromatography-tandem mass spectrometry. *Toxicol. Res.* **2015**, *31*, 255–264. [CrossRef] [PubMed]

42. Wu, C.J.; Wang, L.; Guo, X.B.; Li, H.; Yu, S.J. Simultaneous detection of 4(5)-methylimidazole and acrylamide in biscuit products by isotope-dilution UPLC-MS/MS. *Food Control* **2019**, *105*, 64–70. [CrossRef]

43. Pugajeva, I.; Jaunbergs, J.; Bartkevics, V. Development of a sensitive method for the determination of acrylamide in coffee using high-performance liquid chromatography coupled to a hybrid quadrupole orbitrap mass spectrometer. *Food Addit. Contam. A* **2015**, *32*, 170–179. [CrossRef]

44. Omar, M.M.A.; Elbashir, A.A.; Schmitz, O.J. Determination of acrylamide in Sudanese food by high performance liquid chromatography coupled with LTQ Orbitrap mass spectrometry. *Food Chem.* **2015**, *176*, 342–349. [CrossRef] [PubMed]

45. Arabi, M.; Ghaedi, M.; Ostovan, A. Development of dummy molecularly imprinted based on functionalized silica nanoparticles for determination of acrylamide in processed food by matrix solid phase dispersion. *Food Chem.* **2016**, *210*, 78–84. [CrossRef]

46. Chang, T.T.; Yan, X.Y.; Liu, S.M.; Liu, Y.X. Magnetic dummy template silica sol-gel molecularly imprinted polymer nanospheres as magnetic solid-phase extraction material for the selective and sensitive determination of bisphenol A in plastic bottled beverages. *Food Anal. Method* **2017**, *10*, 3980–3990. [CrossRef]

47. Nasir, A.N.M.; Yahaya, N.; Zain, N.N.M.; Lim, V.; Kamaruzaman, S.; Saad, B.; Nishiyama, N.; Yoshida, N.; Hirota, Y. Thiol-functionalized magnetic carbon nanotubes for magnetic micro-solid phase extraction of sulfonamide antibiotics from milks and commercial chicken meat products. *Food Chem.* **2019**, *276*, 458–466. [CrossRef] [PubMed]

48. Liu, J.M.; Lv, S.W.; Yuan, X.Y.; Liu, H.L.; Wang, S. Facile construction of magnetic core-shell covalent organic frameworks as efficient solid-phase extraction adsorbents for highly sensitive determination of sulfonamide residues against complex food sample matrices. *RSC Adv.* **2019**, *9*, 14247–14253. [CrossRef]

49. Nodeh, H.R.; Ibrahim, W.A.W.; Kamboh, M.A.; Sanagi, M.M. Magnetic graphene sol-gel hybrid as clean-up adsorbent for acrylamide analysis in food samples prior to GC-MS. *Food Chem.* **2018**, *239*, 208–216. [CrossRef] [PubMed]

50. Bagheri, A.R.; Arabi, M.; Ghaedi, M.; Ostovan, A.; Wang, X.Y.; Li, J.H.; Chen, L.X. Dummy molecularly imprinted polymers based on a green synthesis strategy for magnetic solid-phase extraction of acrylamide in food samples. *Talanta* **2019**, *195*, 390–400. [CrossRef] [PubMed]

51. Pourmand, E.; Ghaemi, E.; Alizadeh, N. Determination of acrylamide in potato-based foods using headspace solid-phase microextraction based on nanostructured polypyrrole fiber coupled with ion mobility spectrometry: A heat treatment study. *Anal. Method* **2017**, *9*, 5127–5134. [CrossRef]

52. Nematollahi, A.; Kamankesh, M.; Hosseini, H.; Ghasemi, J.; Hosseini-Esfahani, F.; Mohammadi, A. Investigation and determination of acrylamide in the main group of cereal products using advanced microextraction method coupled with gas chromatography-mass spectrometry. *J. Cereal. Sci.* **2019**, *87*, 157–164. [CrossRef]

53. Wawrzyniak, R.; Jasiewicz, B. Straightforward and rapid determination of acrylamide in coffee beans by means of HS-SPME/GC-MS. *Food Chem.* **2019**, *301*, 125264. [CrossRef]

54. Commission Recommendation (EU). 2013/647 of 8 November 2013 on investigations into the levels of acrylamide in food. *Off. J. Eur. Union* **2013**, *L301*, 15–17.

55. Zokaei, M.; Abedi, A.S.; Kamankesh, M.; Shojaee-Aliababadi, S.; Mohammadi, A. Ultrasonic-assisted extraction and dispersive liquid-liquid microextraction combined with gas chromatography-mass spectrometry as an efficient and sensitive method for determining of acrylamide in potato chips samples. *Food Chem.* **2017**, *234*, 55–61. [CrossRef] [PubMed]

56. Faraji, M.; Hamdamali, M.; Aryanasab, F.; Shabanian, M. 2-Naphthalenthiol derivatization followed by dispersive liquid-liquid microextraction as an efficient and sensitive method for determination of acrylamide in bread and biscuit samples using high-performance liquid chromatography. *J. Chromatogr. A* **2018**, *1558*, 14–20. [CrossRef] [PubMed]

57. Elahi, M.; Kamankesh, M.; Mohammadi, A.; Jazaeri, S. Acrylamide in cookie samples: Analysis using an efficient co-derivatization coupled with sensitive microextraction method followed by gas chromatography-mass spectrometry. *Food Anal. Method* **2019**, *12*, 1439–1447. [CrossRef]

58. Lim, H.H.; Shin, H.S. Ultra trace level determinations of acrylamide in surface and drinking water by GC-MS after derivatization with xanthydrol. *J. Sep. Sci.* **2013**, *36*, 3059–3066. [CrossRef] [PubMed]

59. Backe, W.J.; Yingling, V.; Johnson, T. The determination of acrylamide in environmental and drinking waters by large-volume injection-hydrophilic-interaction liquid chromatography and tandem mass spectrometry. *J. Chromatogr. A* **2014**, *1334*, 72–78. [CrossRef] [PubMed]

60. Wang, H.; Lee, A.W.; Shuang, S.; Choi, M.M. SPE/HPLC/UV studies on acrylamide in deep-fried flour-based indigenous Chinese foods. *Microchem. J.* **2008**, *89*, 90–97. [CrossRef]

61. GB/T 5009. *204–2014. Determination of Acrylamide in Food*; Standard Press: Beijing, China, 2015.

62. Lim, H.H.; Shin, H.S. A new derivatization approach with d-cysteine for the sensitive and simple analysis of acrylamide in foods by liquid chromatography-tandem mass spectrometry. *J. Chromatogr. A* **2014**, *1361*, 117–124. [CrossRef]

63. Ferrer-Aguirre, A.; Romero-Gonzalez, R.; Vidal, J.L.M.; Frenich, A.G. Simple and fast determination of acrylamide and metabolites in potato chips and grilled asparagus by liquid chromatography coupled to mass spectrometry. *Food Anal. Method* **2016**, *9*, 1237–1245. [CrossRef]

64. Yoshioka, T.; Izumi, Y.; Nagatomi, Y.; Miyamoto, Y.; Suzuki, K.; Bamba, T. A highly sensitive determination method for acrylamide in beverages, grains, and confectioneries by supercritical fluid chromatography tandem mass spectrometry. *Food Chem.* **2019**, *294*, 486–492. [CrossRef]

65. Başkan, S.; Erim, F.B. NACE for the analysis of acrylamide in food. *Electrophoresis* **2007**, *28*, 4108–4113. [CrossRef] [PubMed]

66. Voeten, R.L.C.; Ventouri, I.K.; Haselberg, R.; Somsen, G.W. Capillary electrophoresis: Trends and recent advances. *Anal. Chem.* **2018**, *90*, 1464–1481. [CrossRef]

67. Abd El-Hady, D.; Albishri, H.M. Simultaneous determination of acrylamide, asparagine and glucose in food using short chain methyl imidazolium ionic liquid based ultrasonic assisted extraction coupled with analyte focusing by ionic liquid micelle collapse capillary electrophoresis. *Food Chem.* **2015**, *188*, 551–558. [CrossRef] [PubMed]

68. Yang, S.P.; Li, Y.T.; Li, F.; Yang, Z.Y.; Quan, F.F.; Zhou, L.; Pu, Q.S. Thiol-ene click derivatization for the determination of acrylamide in potato products by capillary electrophoresis with capacitively coupled contactless conductivity detection. *J. Agric. Food Chem.* **2019**, *67*, 8053–8060. [CrossRef]

69. Wu, M.L.; Chen, W.J.; Wang, G.; He, P.G.; Wang, Q.J. Analysis of acrylamide in food products by microchip electrophoresis with on-line multiple-preconcentration techniques. *Food Chem.* **2016**, *209*, 154–161. [CrossRef]

70. Kitagawa, F.; Kawai, T.; Sueyoshi, K.; Otsuka, K. Recent progress of on-line sample preconcentration techniques in microchip electrophoresis. *Anal. Sci.* **2012**, *28*, 85–93. [CrossRef]

71. Arvanitoyannis, I.S.; Dionisopoulou, N. Acrylamide: Formation, occurrence in food products, detection methods, and legislation. *Crit. Rev. Food Sci.* **2014**, *54*, 708–733. [CrossRef]

72. Sueyoshi, K.; Kitagawa, F.; Otsuka, K. Effect of a low-conductivity zone on field-amplified sample stacking in microchip micellar electrokinetic chromatography. *Anal. Sci.* **2013**, *29*, 133–138. [CrossRef]

73. Breadmore, M.C.; Tubaon, R.M.; Shallan, A.I.; Phung, S.C.; Keyon, A.S.A.; Gstoettenmayr, D.; Prapatpong, P. Recent advances in enhancing the sensitivity of electrophoresis and electrochromatography in capillaries and microchips (2012–2014). *Electrophoresis* **2015**, *36*, 36–61. [CrossRef]

74. Cui, Y.L.; Zhao, J.; Zhou, J.; Tan, G.Y.; Zhao, Q.Y.; Zhang, Y.H.; Wang, B.M.; Jiao, B.N. Development of a sensitive monoclonal antibody-based indirect competitive enzyme-linked immunosorbent assay for analysing nobiletin in citrus and herb samples. *Food Chem.* **2019**, *293*, 144–150. [CrossRef]

75. Chen, Y.J.; Zhang, S.P.; Hong, Z.S.; Lin, Y.Y.; Dai, H. A mimotope peptide-based dual-signal readout competitive enzyme-linked immunoassay for non-toxic detection of zearalenone. *J. Mater. Chem. B* **2019**, *7*, 6972–6980. [CrossRef] [PubMed]

76. Zhou, X.C.; Shi, J.; Zhang, J.; Zhao, K.; Deng, A.P.; Li, J.G. Multiple signal amplification chemiluminescence immunoassay for chloramphenicol using functionalized SiO_2 nanoparticles as probes and resin beads as carriers. *Spectrochim. Acta. A* **2019**, *222*, 117177. [CrossRef] [PubMed]

77. Li, J.; Zhao, X.; Chen, L.J.; Qian, H.L.; Wang, W.L.; Yang, C.; Yan, X.P. p-Bromophenol-enhanced bienzymatic chemiluminescence competitive immunoassay for ultrasensitive determination of aflatoxin B-1. *Anal. Chem.* **2019**, *91*, 13191–13197. [CrossRef] [PubMed]

78. Chen, E.J.; Xu, Y.; Ma, B.; Cui, H.F.; Sun, C.X.; Zhang, M.Z. Carboxyl-functionalized, europium nanoparticle-based fluorescent immunochromatographic assay for sensitive detection of citrinin in monascus fermented food. *Toxins* **2019**, *11*, 605. [CrossRef]

79. Xiong, Y.; Zhang, K.K.; Gao, B.; Wu, Y.Q.; Huang, X.L.; Lai, W.H.; Xiong, Y.H.; Liu, Y. Fluorescence immunoassay through histone-ds-poly(AT)-templated copper nanoparticles as signal transductors for the sensitive detection of salmonella choleraesuis in milk. *J. Dairy Sci.* **2019**, *102*, 6047–6055. [CrossRef]

80. Singh, G.; Brady, B.; Koerner, T.; Becalski, A.; Zhao, T.; Feng, S.; Godefroy, S.B.; Huet, A.C.; Delahaut, P. Development of a highly sensitive competitive indirect enzyme-linked immunosorbent assay for detection of acrylamide in foods and water. *Food Anal. Method* **2014**, *7*, 1298–1304. [CrossRef]

81. Wu, J.; Shen, Y.D.; Lei, H.T.; Sun, Y.M.; Yang, J.Y.; Xiao, Z.L.; Wang, H.; Xu, Z.L. Hapten synthesis and development of a competitive indirect enzyme-linked immunosorbent assay for acrylamide in food samples. *J. Agr. Food Chem.* **2014**, *62*, 7078–7084. [CrossRef]

82. Zhu, Y.T.; Song, S.S.; Liu, L.Q.; Kuang, H.; Xu, C.L. An indirect competitive enzyme-linked immunosorbent assay for acrylamide detection based on a monoclonal antibody. *Food Agric. Immunol.* **2016**, *27*, 796–805. [CrossRef]

83. Assaat, L.D.; Saepudin, E.; Soejoedono, R.D.; Adji, R.S.; Poetri, O.N.; Ivandini, T.A. Production of a polyclonal antibody against acrylamide for immunochromatographic detection of acrylamide using strip tests. *J. Anim. Vet. Adv.* **2019**, *6*, 366–375. [CrossRef]

84. Sun, Q.; Xu, L.H.; Ma, Y.; Qiao, X.G.; Xu, Z.X. Study on a biomimetic enzyme-linked immunosorbent assay method for rapid determination of trace acrylamide in French fries and cracker samples. *J. Sci. Food Agric.* **2014**, *94*, 102–108. [CrossRef]

85. Sarkar, D.; Xie, X.J.; Anselmo, A.C.; Mitragotri, S.; Banerjee, K. MoS_2 Field-effect transistor for next-generation label-free biosensors. *ACS Nano.* **2014**, *8*, 3992–4003. [CrossRef]

86. Hu, Q.Q.; Wang, R.H.; Wang, H.; Slavik, M.F.; Li, Y.B. Selection of acrylamide-specific aptamers by a quartz crystal microbalance combined SELEX method and their application in rapid and specific detection of acrylamide. *Sens. Actuators B Chem.* **2018**, *273*, 220–227. [CrossRef]

87. Bhadani, S.N.; Prasad, Y.K.; Kundu, S. Electrochemical and chemical polymerization of acrylamide. *J. Polym. Sci. Pol. Chem.* **1980**, *18*, 1459–1469. [CrossRef]

88. Casella, I.G.; Pierri, M.; Contursi, M. Determination of acrylamide and acrylic acid by isocratic liquid chromatography with pulsed electrochemical detection. *J. Chromatogr. A* **2006**, *1107*, 198–203. [CrossRef]

89. Trojanowicz, M.; Hitchman, M.L. Determination of pesticides using electrochemical biosensors. *Trac-Trends Anal. Chem.* **2015**, *14*, 1311–1328. [CrossRef]

90. Wu, D.; Du, D.; Lin, Y. Recent progress on nanomaterial-based biosensors for veterinary drug residues in animal-derived food. *Trends Anal. Chem.* **2016**, *83*, 95–101. [CrossRef]

91. Pandey, C.M.; Malhotra, B.D. *Nanomaterials for Biosensors*; Biosensors: Basel, Switzerland, 2018; pp. 1–74.

92. Rasooly, A.; Herold, K.E. Biosensors for the analysis of food-and waterborne pathogens and their toxins. *J. Aoac. Int.* **2019**, *89*, 873–883. [CrossRef]

93. Shiddiky, M.J.A.; Torriero, A.A.J. Application of ionic liquids in electrochemical sensing systems. *Biosens. Bioelectron.* **2011**, *26*, 1775–1787. [CrossRef]

94. Batra, B.; Lata, S.; Pundir, C.S. Construction of an improved amperometric acrylamide biosensor based on hemoglobin immobilized onto carboxylated multi-walled carbon nanotubes/iron oxide nanoparticles/chitosan composite film. *Bioproc. Biosyst. Eng.* **2013**, *36*, 1591–1599. [CrossRef]

95. Garabagiu, S.; Mihailescu, G. Simple hemoglobin-gold nanoparticles modified electrode for the amperometric detection of acrylamide. *J. Electroanal. Chem.* **2011**, *659*, 196–200. [CrossRef]

96. Sayyad, A.S.; Balakrishnan, K.; Ci, L.; Kabbani, A.T.; Vajtai, R.; Ajayan, P.M. Synthesis of iron nanoparticles from hemoglobin and myoglobin. *Nanotechnology* **2012**, *23*, 1–5. [CrossRef]

97. Wu, H.; Wang, X.; Qiao, M. Enhancing sensitivity of hemoglobin-based electrochemical biosensor by using protein conformational intermediate. *Sens. Actuators B-Chem.* **2015**, *22*, 1694–1699. [CrossRef]

98. Yadav, N.; Chhillar, A.K.; Pundir, C.S. Preparation, characterization and application of haemoglobin nanoparticles for detection of acrylamide in processed foods. *Int. J. Biol. Macromol.* **2018**, *107*, 1000–1013. [CrossRef]

99. Asnaashari, M.; Kenari, R.E.; Farahmandfar, R.; Abnous, K.; Taghdisi, S.M. An electrochemical biosensor based on hemoglobin-oligonucleotides-modified electrode for detection of acrylamide in potato fries. *Food Chem.* **2019**, *271*, 54–61. [CrossRef]

100. Bucur, M.P.; Bucur, B.; Radu, G.L. Simple, selective and fast detection of acrylamide based on glutathione S-transferase. *RSC Adv.* **2018**, *8*, 23931–23936. [CrossRef]

101. Sumner, S.C.; MacNeela, J.P.; Fennell, T.R. Characterization and quantitation of urinary metabolites of [1,2,3-^{13}C]acrylamide in rats and mice using ^{13}C nuclear magnetic resonance spectroscopy. *Chem. Res. Toxicol.* **1992**, *5*, 81–89. [CrossRef]

102. Hartmann, E.C.; Bolt, H.M.; Drexler, H.; Angerer, J. N-Acetyl-S-(1-carbamoyl-2-hydroxy-ethyl)-l-cysteine (iso-GAMA) a further product of human metabolism of acrylamide: Comparison with the simultaneously excreted other mercaptuic acids. *Arch. Toxicol.* **2009**, *83*, 731–734. [CrossRef] [PubMed]

103. Hartmann, E.C.; Latzin, J.M.; Schindler, B.K.; Koch, H.M.; Angerer, J. Excretion of 2,3-dihydroxy-propionamide (OH-PA), the hydrolysis product of glycidamide, in human urine after single oral dose of deuterium-labeled acrylamide. *Arch. Toxicol.* **2011**, *85*, 601–606. [CrossRef]

104. Luo, Y.S.; Long, T.Y.; Shen, L.C.; Huang, S.L.; Chiang, S.Y.; Wu, K.Y. Synthesis, characterization and analysis of the acrylamide- and glycidamide-glutathione conjugates. *Chem. Biol. Interact.* **2015**, *237*, 38–46. [CrossRef]

105. Wulandari, R.; Ivandini, T.A.; Irkham; Saepudin, E.; Einaga, Y. Modification of boron-doped diamond electrodes with platinum to increase the stability and sensitivity of haemoglobin-based acrylamide sensors. *Sens. Mater.* **2019**, *31*, 1105–1117.

106. Wu, M.F.; Wang, Y.; Li, S.; Dong, X.X.; Yang, J.Y.; Shen, Y.D.; Wang, H.; Sun, Y.M.; Lei, H.T.; Xu, Z.L. Ultrasensitive immunosensor for acrylamide based on chitosan/SnO$_2$-SiC hollow sphere nanochains/gold nanomaterial as signal amplification. *Anal. Chim. Acta* **2019**, *1049*, 188–195. [CrossRef] [PubMed]

107. Rivas, G.A.; Rubianes, M.D.; Rodriguez, M.C.; Ferreyra, N.E.; Luque, G.L.; Pedano, M.L.; Miscoria, S.A.; Parrado, C. Carbon nanotubes for electrochemical biosensing. *Talanta* **2007**, *74*, 291–307. [CrossRef]

108. Deshmukh, M.A.; Jeon, J.Y.; Ha, T.J. Carbon nanotubes: An effective platform for biomedical electronics. *Biosens. Bioelectron.* **2020**, *150*, 111919. [CrossRef] [PubMed]

109. Liu, X.; Mao, L.G.; Wang, Y.L.; Shi, X.B.; Liu, Y.; Yang, Y.; He, Z. Electrochemical sensor based on imprinted sol-gel polymer on Au NPs-MWCNTs-CS modified electrode for the determination of acrylamide. *Food Anal. Method.* **2016**, *9*, 114–121. [CrossRef]

110. Batra, B.; Lata, S.; Sharma, M.; Pundir, C.S. An acrylamide biosensor based on immobilization of hemoglobin onto multiwalled carbon nanotube/copper nanoparticles/polyaniline hybrid film. *Anal. Biochem.* **2013**, *433*, 210–217. [CrossRef]

111. Varmira, K.; Abdi, O.; Gholivand, M.B.; Goicoechea, H.C.; Jalalvand, A.R. Intellectual modifying a bare glassy carbon electrode to fabricate a novel and ultrasensitive electrochemical biosensor: Application to determination of acrylamide in food samples. *Talanta* **2018**, *176*, 509–517. [CrossRef]

112. Wolfbeis, O.S. Materials for fluorescence-based optical chemical sensors. *J. Mater. Chem.* **2005**, *15*, 2657–2669. [CrossRef]

113. Dandin, M.; Abshire, P.; Smela, E. Optical filtering technologies for integrated fluorescence sensors. *Lab Chip* **2007**, *7*, 955–977. [CrossRef]

114. Dhenadhayalan, N.; Lee, H.L.; Yadav, K.; Lin, K.C.; Lin, Y.T.; Chang, A.H.H. Silicon quantum dots based fluorescence turn-on metal ion sensors in live cells. *ACS Appl. Mater. Inter.* **2016**, *8*, 3953–23962. [CrossRef]

115. Li, X.; Li, Z.; Yang, Y.W. Tetraphenylethylene-interweaving conjugated macrocycle polymer materials as two-photon fluorescence sensors for metal ions and organic molecules. *Adv. Mater.* **2018**, *30*, 1800177. [CrossRef]

116. Al-Zahrani, F.A.M.; El-Shishtawy, R.M.; Asiri, A.M.; Al-Soliemy, A.M.; Abu Mellah, K.; Ahmed, N.S.E.; Jedidi, A. A new phenothiazine-based selective visual and fluorescent sensor for cyanide. *BMC Chem.* **2020**, *14*, 2. [CrossRef]

117. Kong, Y.L.; Cheng, Q.; He, Y.; Ge, Y.L.; Zhou, J.G.; Song, G.W. A dual-modal fluorometric and colorimetric nanoprobe based on graphitic carbon nitrite quantum dots and Fe(II)-bathophenanthroline complex for detection of nitrite in sausage and water. *Food Chem.* **2020**, *312*, 126089. [CrossRef]

118. Huang, B.H.; Shen, S.S.; Wei, N.; Guo, X.F.; Wang, H. Fluorescence biosensor based on silicon quantum dots and 5,5′-dithiobis-(2-nitrobenzoic acid) for thiols in living cells. *Spectrochim. Acta A* **2020**, *229*, 117972. [CrossRef]

119. Zhou, Y.; Ma, Z.F. A novel fluorescence enhanced route to detect copper(II) by click chemistry-catalyzed connection of Au@SiO$_2$ and carbon dots. *Sens. Actuators B-Chem.* **2016**, *233*, 426–430. [CrossRef]

120. Mani, N.P.; Ganiga, M.; Cyriac, J. MoS$_2$ nanohybrid as a fluorescence sensor for highly selective detection of dopamine. *Analyst* **2018**, *143*, 1691–1698. [CrossRef]

121. Peng, D.; Zhang, L.; Liang, R.P.; Qiu, J.D. Rapid detection of mercury ions based on nitrogen doped graphene quantum dots accelerating formation of manganese porphyrin. *ACS Sens.* **2018**, *3*, 1040–1047. [CrossRef]

122. Hu, Q.Q.; Xu, X.H.; Li, Z.M.; Zhang, Y.; Wang, J.P.; Fu, Y.C.; Li, Y.B. Detection of acrylamide in potato chips using a fluorescent sensing method based on acrylamide polymerization-induced distance increase between quantum dots. *Biosens. Bioelectron.* **2014**, *54*, 64–71. [CrossRef]

123. Liu, Y.; Hu, X.; Bai, L.; Jiang, Y.H.; Qiu, J.; Meng, M.J.; Liu, Z.C.; Ni, L. A molecularly imprinted polymer placed on the surface of graphene oxide and doped with Mn(II)-doped ZnS quantum dots for selective fluorometric determination of acrylamide. *Microchim. Acta* **2018**, *185*, UNSP 48. [CrossRef]

124. Asnaashari, M.; Kenari, R.E.; Farahmandfar, R.; Taghdisi, S.M.; Abnous, K. Fluorescence quenching biosensor for acrylamide detection in food products based on double-stranded DNA and gold nanoparticles. *Sens. Actuators B-Chem.* **2018**, *265*, 339–345. [CrossRef]

125. Amit, N.; Anupam, B.; Shweta, P. Gold nanoparticle induced enhancement of molecular fluorescence for Zn2+ detection in aqueous niosome solution. In Proceedings of the 13th International Conference on Fibre Optics & Photonics, Kanpur, India, 4–8 December 2016; Optical Society of America: Washington, DC, USA, 2016; Paper W2D.1.

 *foods*

Article

Determination of Acrylamide in Biscuits by High-Resolution Orbitrap Mass Spectrometry: A Novel Application

Cristiana L. Fernandes, Daniel O. Carvalho and Luis F. Guido *

REQUIMTE—Departamento de Química e Bioquímica, Faculdade de Ciências, Universidade do Porto, Rua do Campo Alegre 687, 4169-007 Porto, Portugal; cris.rlf@gmail.com (C.L.F.); daniel.carvalho@fc.up.pt (D.O.C.)
* Correspondence: lfguido@fc.up.pt; Tel.: +351-220-402-644

Received: 14 October 2019; Accepted: 18 November 2019; Published: 20 November 2019

Abstract: Acrylamide (AA), a molecule which potentially increases the risk of developing cancer, is easily formed in food rich in carbohydrates, such as biscuits, wafers, and breakfast cereals, at temperatures above 120 °C. Thus, the need to detect and quantify the AA content in processed foodstuffs is eminent, in order to delineate the limits and mitigation strategies. This work reports the development and validation of a high-resolution mass spectrometry-based methodology for identification and quantification of AA in specific food matrices of biscuits, by using LC-MS with electrospray ionization and Orbitrap as the mass analyser. The developed analytical method showed good repeatability (RSD_r 11.1%) and 3.55 and 11.8 µg kg^{-1} as limit of detection (LOD) and limit of quantification (LOQ), respectively. The choice of multiplexed targeted-SIM mode (t-SIM) for AA and AA-d3 isolated ions provided enhanced detection sensitivity, as demonstrated in this work. Statistical processing of data was performed in order to compare the AA levels with several production parameters, such as time/cooking temperature, placement on the cooking conveyor belt, color, and moisture for different biscuits. The composition of the raw materials was statistically the most correlated factor with the AA content when all samples are considered. The statistical treatment presented herein enables an important prediction of factors influencing AA formation in biscuits contributing to putting in place effective mitigation strategies.

Keywords: acrylamide; biscuits; mitigation measures; benchmark levels; contaminant

---

1. Introduction

Once Tareke et al. [1] have reported acrylamide (AA) as a carcinogen formed in heated foodstuffs in the food industry, Member States of the European Union and the European Commission have made considerable efforts to investigate AA formation pathways in order to reduce the levels of this compound in processed foods. In addition to being present in foods, AA has also been found in the environment (due to industrial discharges), cosmetics, drinking water, as well as tobacco smoke. Human exposure to AA may be by ingestion, inhalation, or contact with the skin [2]. Dietary exposure is the most concerning, since acrylamide is present in a wide range of everyday foods. Between 10% and 50% of AA of the diet of pregnant women passes through the placenta and breast milk also contains this compound [3]. In the US, most exposure to AA comes from potato chips, breads, cereals, crackers, and other snacks [4]. In Europe, toasted bread, coffee, and potatoes are the main food sources of AA [5].

Human exposure to AA may have toxicological effects (neurotoxicity, genotoxicity, carcinogenicity, and reproductive toxicity), and AA has been classified as carcinogenic by the International Agency for Research on Cancer [6] in the 2A group (probably carcinogenic in humans). AA has an α, β-unsaturated carbonyl group with electrophilic reactivity, which can react with nucleophilic groups of biological molecules, thus contributing to toxic effects. The reaction of AA with proteins is extensive and the products of this reaction are used as biomarkers of its presence [5]. It is metabolized together with glutathione (GSH) and also by epoxidation, resulting in glycidamide (GA). The formation of GA is mediated preferentially by cytochrome P450, and is on the basis of neuro and genotoxicity of AA. Covalent DNA adducts of GA were observed in vitro and in animal experiments and were used as biomarkers [2,5].

Only the legal limit of AA for water has been established, with the value of 0.1 $\mu g\ L^{-1}$ [7]. The levels of AA in foodstuffs of the Member States of the European Union were monitored between 2007 and 2012. Based on the results, the European Commission outlined indicative values for AA in different foodstuffs [8]. According to Regulation 2017/2158, these values are not safety values but rather indicative values, so that further research is promoted in foods with higher AA levels and consequent reduction throughout the agronomical factors, the food recipe, processing, and final preparation [5].

The level of free asparagine in cereals has been claimed to be the major influence on the formation of AA [9], since the largest pathway of AA formation involves this amino acid. The choice of cereal varieties with lower levels of free asparagine is recommended, but challenging given the influence of environmental conditions on their production [10].

Corn and rice products tend to have lower AA contents than wheat, barley, oats, or rye. Products with whole flours have higher levels of AA [10]. The choice of different varieties of cereals also determines the development of AA: Five varieties of rye with different fertilizations were used to study the effect of nitrogen and sulfur on AA formation. A positive correlation was found between asparagine concentration in grains and the highest levels of nitrogen used and the final concentration of AA [11,12].

The influence of cereal types on bread was investigated by Przygodzka et al. [13], concluding that rye loaves form more AA in cooking, followed by spelled loaves and loaves of refined flour—"white bread." In the same study, the extraction rates of the flour were compared with AA formation: 100% whole flours obtained higher concentrations of AA, followed by flours with extractions of 70%, indicating that "whole flours" have more AA precursors.

Several AA mitigation measures have been established that involve the use of the enzyme asparaginase, which converts asparagine to aspartic acid, although control of adverse effects on organoleptic properties is necessary [14].

The requirement for ultra-trace level detection of AA has led to the development of several analytical methods, most of which involve chromatographic separation techniques, both liquid and gas chromatography. Determination of AA in food by GC-MS methods can be carried out with or without derivatization. The advantage of derivatization processes is increased volatility and improved selectivity. The bromination [1,15–20], xanthydrol [21–25], and silylation [26,27] have been widely used for determining of AA in foodstuffs.

In recent years, the use of ultra-performance liquid chromatography (UPLC) has become more popular because of its high sensitivity and selectivity, without the need for derivatization. Liquid chromatography coupled to mass spectrometry have become the method of choice for the determination of AA in food products, by using different mass analysers. Conventional triple quadrupole (QqQ) have been for long the technique of choice by selecting the characteristic transitions m/z 72→55, and 72→27. Ion trap [28–30] and TOF [28,31–33] have also been useful for quantitative analyses of AA.

Considering that the capabilities of high-resolution mass spectrometry (HRMS) based methodologies for quantitative LC/MS analysis of AA in foodstuffs have been scarcely explored [34–36] the present work aims at validating a HRMS methodology for detection and quantification of AA in biscuits. The implemented procedure has been applied for investigating the impact of several production parameters on the AA content in biscuits.

2. Materials and Methods

2.1. Chemicals

Acrylamide ($\geq$ 95% for HPLC) and acrylamide-d3 standard solution (1000 mg L^{-1} in acetonitrile) were purchased from Sigma-Aldrich (Steinheim, Germany). Methanol (for UHPLC), ethanol (99.5%), and dichloromethane (for HPLC) were from Panreac (Barcelona, Spain). High-purity water from a Millipore Simplicity 185 water purification system (Millipore Iberian S. A., Madrid, Spain) was used for all chemical analyses and glassware washing. The solvents employed for HPLC were filtered through a Nylon filter of 0.45 μm pore size (Whatman, Clifton, NJ, USA) and degasified for 10 min in an ultrasound bath.

2.2. Standard Solutions

Concentrated stock solutions of acrylamide (1 mg mL^{-1}) and acrylamide-d3 (0.5 mg mL^{-1}), used as internal standard, were prepared by dissolving the compounds in ethanol. Diluted standard solutions were further prepared by adding the appropriate volume of each stock solution to water.

2.3. Biscuit Samples

This work has been carried out in close collaboration with a leading company at the national level and with an international dimension, whose confidentiality will be maintained for obvious reasons. Four biscuit types were supplied (hereinafter referred to A, B, C, and D), collected from three different sample points in the baking oven, as depicted in Figure 1. Biscuits A, B, and C are made from wheat flour type 65 whereas Biscuit D is made from wheat bran. In addition, Biscuit C has cocoa in its composition.

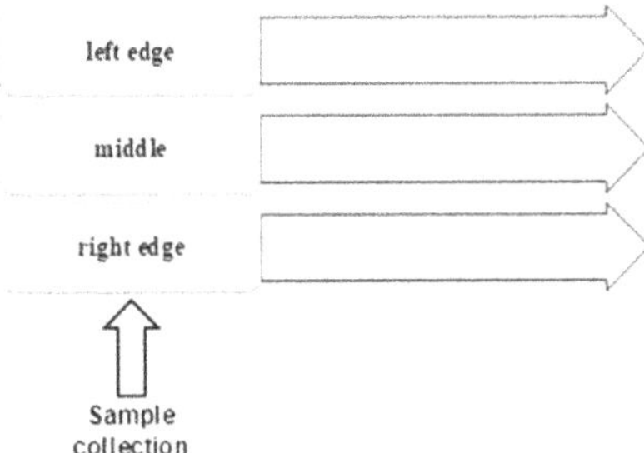

Figure 1. Sample collection points from the baking oven.

2.4. Sample Preparation

Between three and five biscuits were pooled and grinded in a solid sample grinder (Moulinex, France) and put through an Endecott's test sieve (London, England). Approximately 1 g of each ground homogenous sample were transferred into a 50 mL polypropylene graduated conical tube with cap. 250 ng of internal standard (acrylamide-d3) and then 15 mL of ultrapure water were added to each tube, which was placed in the ultrasonic bath for 15 min. 2 mL of dichloromethane was added to each tube, left on the rotary shaker for further 20 min. The tubes were centrifuged at 5000 rpm for 15 min. 1500 μL of supernatant from each tube was withdrawn for extraction and purification by solid phase extraction (SPE).

For the SPE clean-up, the Oasis HLB SPE cartridge (6 mL, 200 mg, 30 μm particle size from Waters) was conditioned under vacuum with methanol (3.5 mL), and equilibrated with water (3.5 mL). Then, 1.5 mL of the withdrawn supernatant were loaded on the Oasis HLB SPE cartridge and allowed to pass completely through the sorbent material. The cartridge was rinsed with 500 μL of ultrapure water and samples were eluted with 1.5 mL of water.

For the second step of the clean-up, the Bond Elut AccuCAT SPE cartridge (3 mL, 200 mg, 50 μm particle size from Agilent Technologies) was conditioned under vacuum with methanol (2.5 mL), and equilibrated with water (2.5 mL). Then, the cartridge was loaded with the solution from the previous step and 1 mL was discarded. The remaining volume was collected directly to an injection vial.

2.5. LC-ESI-Orbitrap

The samples were separated on Accela HPLC (Thermo Fischer Scientific, Bremen, Germany) Electrospray Orbitrap, using a C18 Phenomenex Germini (Phenomenex, USA), particle size of three microns and size 4.6 mm ID × 150 mm. The samples were eluted through a gradient of 90% solvent A (0.1% HCOOH in water) and 10% solvent B (methanol) for 2 min at a flow rate of 0.4 mL/min, thereafter for 18 min over 100% solvent B and 10 min in a 10% solvent B gradient.

The analysis was performed on a hybrid mass spectrometer LTQ XL OrbitrapTM (Thermo Fischer Scientific, Bremen, Germany), controlled by LTQ Tune Plus Xcalibur 2.5.5 and 2.1.0. The following ionisation (positive mode) parameters were applied: Electrospray voltage 3.2 kV, capillary temperature 300 °C, sheath gas (N_2), 40 arbitrary units (arb), auxiliary gas (N_2) 10 (arb), and S-Lens RF level at 25 (arb). The automatic gain control was used to fill the C-trap and gain accuracy in mass measurements (ultimate mass accuracy mode, 1×10^5 ions), the SIM maximum IT was set to 50 ms, the number of micro-scans to be performed was set at three. Mass spectra were recorded in multiplexed targeted-SIM mode (t-SIM) with a mass resolving power of 60,000 full width at half maximum (FWHM) with a quadrupole isolation window of 1.0 Da for isolated ions (72.0444 Da for acrylamide and 75.0632 Da for acrylamide-d3). Chromatograms for a biscuit sample, indicating the acrylamide (AA) and deuterated acrylamide (acrylamide-d3) retention times, are shown in Figure S1.

2.6. Analysis of Colour

The biscuits' colour (Biscuit D) was analyzed with the Minolta CR-410 colorimeter. The parameters used were Luminosity (*L*), Red (*a*), and Yellow (*b*). The biscuits were analyzed in their form of consumption (without being ground), so that the color could be considered a method of control in future industrial tests and quality parameters.

2.7. Moisture Content Determination

The biscuits' moisture level (Biscuit D) was assessed on the same day of the AA extraction. About 5 g of ground biscuit were dried for 3 h at 100 °C. After drying and cooling, the dry mass of the biscuit was measured and the moisture content was calculated.

2.8. Statistical Analysis

To measure the strength of relationship between the measured variables, Pearson's correlation coefficient (*r*) and Spearman's correlation coefficient (*ρ*) were calculated. While the Pearson correlation coefficient reflects the strength of linear relationships, the Spearman rank correlation reflects the strength of monotonic relationship [37].

The statistical package StatBox 7.5 (Grimmer Logiciel, Paris, France) was used for all statistical calculations.

3. Results and Discussion

3.1. Method Performance

Limits of detection and quantification (LOD and LOQ) were estimated by using the signal-to-noise method, as specified in the European Pharmacopoeia [38]. The peak-to-peak noise around the AA (*m/z* 72.0444) retention time was measured, and subsequently, the concentration of the AA that yielded a signal equal to a certain value of noise to signal ratio was estimated, by comparing measured signals from samples with known low concentrations of the AA with those of blank samples. This method allows a decrease of the signal (peak height) to be observed to the extent that the concentration is reduced through a series of dilutions, establishing the minimum concentration at which the analyte can be reliably quantified. The signal-to-noise (S/N) ratios accepted as estimates of the LOD and LOQ were 3:1 and 10:1, respectively [39]. The values found in this study, based on three measurements of Biscuit A, are 3.55 µg kg^{-1} for LOD and 11.8 µg kg^{-1} for LOQ, as shown in Table 1.

Table 1. Repeatability, limit of detection (LOD), and limit of quantification (LOQ) of the proposed methodology, based on several measurements of Biscuit A.

Assay	Acrylamide (AA) Content (µg kg^{-1})	Average AA Content (µg kg^{-1})	SD	RSD$_r$ %	LOD (µg kg^{-1})	LOQ (µg kg^{-1})
1	254.1					
2	277.3					
3	343.9					
4	309.8					
5	269.8	297.9	33.1	11.1	3.55	11.8
6	345.0					
7	290.9					
8	292.0					

Commission regulation of 20 November 2017 states that the method of analysis used for the analysis of AA must comply with the following criteria: LOQ less than or equal to two fifths of the benchmark level (for benchmark level <125 µg kg^{-1}) and less than or equal to 50 µg kg^{-1} (for benchmark level ≥125 µg kg^{-1}); LOD less than or equal to three tenths of LOQ [40]. According to the same regulation, the benchmark level for the presence of AA in biscuits and wafers is 350 µg kg^{-1}. This means that LOD and LOQ is required to be less than or equal to 15 and 50 µg kg^{-1}, respectively. The method herein presented clearly meets these requirements. Moreover, in a proficiency test recently organized by the EURL-PAH, for the determination of the AA content in potato chips, the method performance LOD and LOQ were reported [41]. Twenty-six laboratories guarantee the determination of AA with an average LOD of 22.5 µg kg^{-1} and LOQ 55.8 µg kg^{-1} by liquid chromatography (LC) coupled with mass spectrometry (MS; MS/MS). Nine laboratories participating in this proficiency test reported 15.2 and 36.3 µg kg^{-1} as average LOD and LOQ, respectively, based on GC-MS methods. By using the analytical method herein reported, it is possible to increase the detectability and thus achieve lower limit of quantification, which can be particularly useful in the case of low-abundance AA matrices. The selection of the acquisition mode in the Orbitrap has a direct impact on the detection sensitivity. In a recent paper, Kaufmann demonstrated that the sensitivity of eight selected analytes is strongly increased by the use of SIM (selected ion monitoring) relatively to the FS (full scan) mode (1.5-fold increase for analytes in pure standard solutions and 2-fold increase for analytes spiked in a heavy matrix) [42]. A detailed study of the acquisition method for determination of eight synthetic hormones in animal urine concluded that reducing the scan range for Full MS (using the quadrupole) and targeted modes give higher S/N ratios and thereby better detection limits for analytes in complex matrices [43]. In fact, the targeted-SIM (t-SIM) is not more selective than full MS, but it does provide enhanced detection sensitivity. As only a small fraction of the continuously entering ion beam is sampled by the C-trap, the number of ions transmitted is greatly reduced and a much longer segment of the ion beam can be collected. Accordingly, significantly higher sensitivity can be achieved, mainly for small molecules applications, such as the present case of AA.

The precision of the method was evaluated by measuring the repeatability (intra-day variability). The relative standard deviation was calculated for repeatability (RSDr) by performing eight repeated analyses for samples of the same biscuit. The results showed that the RSDr (11.1%) was less than 12% for a sample with an average AA content of 297.9 µg kg^{-1} (Table 1). The use of isotopically labeled internal standard (acrylamide-d3) is herein especially useful, as sample loss may occur during sample preparation steps prior to analysis, as it is known that the fat/water distribution of the matrix may affect the extraction and analysis.

3.2. Acrylamide Content in Biscuits

The optimized and validated procedure was applied to different samples of biscuits collected from the baking oven. Three sample points were considered, as depicted in Figure 1. One is in the middle of the oven and two are in the edges of the oven (left edges and right edges).

A considerable difference was observed between samples collected from the middle and edges of the oven (Table 2). Except for Biscuit B, the AA content is higher for samples taken from the middle of the oven, where the temperatures are higher. The observed increase is higher for Biscuits D (from 1443 up to 3303 µg kg^{-1}, corresponding to 129% increase) and A (from 216 up to 431 µg kg^{-1}, corresponding to 99% increase). Except for Biscuit A, the AA content found for all the inspected biscuits was above the benchmark level referred to in the European Union (EU) Commission Regulation [40]. Of more concern is the fact that Biscuits C and D contain AA in concentration clearly above (average 2056 and 2373 µg kg^{-1}, respectively) the indicative value reported by European Food Safe Authority (dashed line in Figure 2), confirming the pressure of establishing mitigation measures for the reduction of the presence of AA in these matrices.

Table 2. Acrylamide content (µg kg^{-1}) in biscuits collected from different points of the baking oven.

Biscuit	Edges of the Baking Oven (µg kg^{-1})	Middle of the Baking Oven (µg kg^{-1})	Average (µg kg^{-1})
A	216	431	324 ± 36
B	563	551	557 ± 61
C	1881	2231	2056 ± 226
D	1443	3303	2373 ± 261

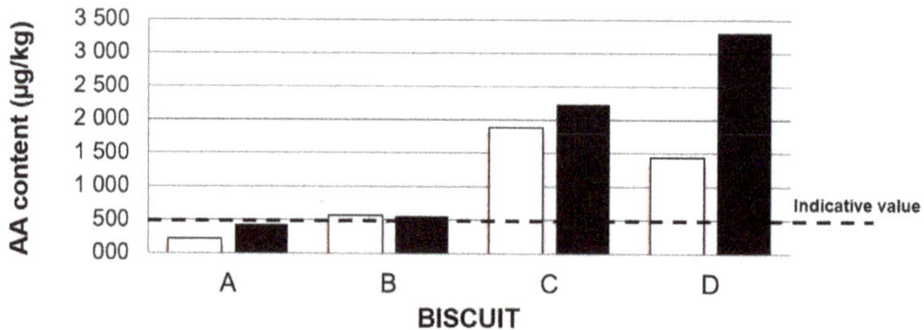

Figure 2. Effect of the position in the oven on the acrylamide content of biscuits. The dashed line depicts the indicative level (500 µg kg^{-1}) reported by the European Food Safe Authority [8]. □ edges of the baking oven ■ middle of the baking oven.

The current analyses are in line with the hypothesis that the raw materials are the major factors influencing the formation of AA, in particular the asparagine content of cereal flours [44–47]. The highest value obtained for Biscuit D (average 2373 µg kg^{-1}) can be justified by its composition, since one of its raw materials is the wheat bran. Wheat bran is the outer part of the wheat grain, which is removed in flours such as wheat flour type 65. "Whole" flour contains wheat bran and are associated with higher concentrations of asparagine (691 mg kg^{-1}) compared to wheat flour type 65 (54.5 mg kg^{-1}). Another type of raw material that may increase the concentration of AA in biscuits is that undergoing heat treatment, such as cocoa. Cocoa, which is a raw material with thermal pretreatment and therefore prone to the formation of AA [48], is present in Biscuit C explaining the high content found (average 2056 µg kg^{-1}).

3.3. Correlation between Acrylamide Content and Biscuit Colour

The AA content has been compared with both the colour, by measuring the Hunter Scale parameters, L, a and b in a colorimeter, as well as with the moisture content of biscuits collected in different oven positions. As can be seen in the images shown in Figure 3, the browning of the biscuit associated with the increase of the AA content is clearly observed. The Hunter colour scale parameters L (light) and a (red) increased during cooking steps (ΔL = 5.53 and 2.25; Δa = 3.61 and 0.61) confirming that a correlation exists between the AA content and the browning of the biscuit.

Figure 3. Photograph of the same lot of biscuits (Biscuit D) subject to different cooking temperatures. The different colors can be observed (ΔL = 5.53 and 2.25; Δa = 3.61 and 0.61), as well as the corresponding acrylamide levels (ΔAA = 1035 and 1860 µg kg^{-1}).

Multivariate statistical analyses were carried out in order to define the parameters which most well correlate with the AA content (Table 3). The L parameter, which is luminosity, is the first variable correlated with AA content whereas an inverse correlation was found between the moisture and the AA content. This is not surprising, since the temperature of the baking oven is expected to be negatively correlated with the final moisture content of the biscuit, showing the direct impact of temperature on the AA content. The influence of temperature on the formation of AA thus seems confirmed, as demonstrated in previous studies [13,49,50]. In addition, Jozinovic et al. [51] have recently shown that the moisture content and temperature during extrusion had a greater impact on the formation of AA in relation to screw speed. Recent results revealed that at low temperatures used for the thermal treatment, the amount of AA formed was lower, even if the treatment duration was longer [52]. In the current work, the baking times are identical for the four biscuit types, thus it is not possible to associate them with the AA values analyzed.

Table 3. Correlation matrix between the acrylamide content and the colour and moisture of biscuits (Biscuit D). In bold, significant values (except diagonal) at the level of significance 95%.

	AA Content	L	a	b	Moisture
AA content	1.0				
L	**0.541**	1.0			
a	0.310	0.352	1.0		
b	0.276	**0.864**	0.663	1.0	
Moisture	**−0.277**	0.307	0.304	0.576	1.0

4. Conclusions

A sensitive and efficient HRMS methodology, based on LC-MS with electrospray ionization and Orbitrap as mass analyser, allowing quantification of AA for specific food matrices of biscuits was presented. Combining the multiplexed targeted-SIM mode for AA and isotopically labeled internal standard (acrylamide-d3), the proposed HRMS method enables reliable and accurate analyses of AA with very little influence by the matrix components. Under these conditions 3.55 μg kg^{-1} for LOD and 11.8 μg kg^{-1} for LOQ are attainable.

During baking an increase in AA concentration was observed, as well as for samples taken from the middle of the oven, where the temperatures are higher. Statistical processing of data shows that the composition of the raw materials of the biscuits was statistically the most correlated factor with the AA content. Statistical treatment shows the direct impact of temperature on the AA content as well.

This study also reported that two types of biscuits (out of four) contain AA in concentration clearly above the indicative value reported by European Food Safe Authority, confirming the pressure of establishing mitigation measures for the reduction of the presence of AA in these matrices.

Supplementary Materials: The following are available online at http://www.mdpi.com/2304-8158/8/12/597/s1, Figure S1: Chromatograms for a biscuit sample, indicating the acrylamide (AA) and deuterated acrylamide (AA-d3) retention times.

Author Contributions: Conceptualization, L.F.G.; Methodology, C.L.F. and D.O.C.; Data curation, C.L.F. and L.F.G.; Writing—original draft preparation, L.F.G.; Writing—review and editing, all authors; Supervision, L.F.G.

Funding: This research received financial support from the European Union (FEDER funds POCI/01/0145/ FEDER/007265) and from FCT/MEC through national funds and co-financed by FEDER (UID/QUI/50006/2013-NORTE-01-0145-FEDER-00011) under the Partnership Agreement PT2020. DOC receives a postdoc grant through the project Operação NORTE-01-0145-FEDER-000011. Mass spectrometric analyses were conducted at CEMUP (Materials Centre of the University of Porto, Portugal) supported by the project NORTE-07-0162-FEDER-00048.

Acknowledgments: The authors are thankful to Silvia Maia (CEMUP) for her technical assistance in the mass spectrometric analyses.

Conflicts of Interest: All authors disclose any potential sources of conflict of interest.

References

1. Tareke, E.; Rydberg, P.; Karlsson, P.; Eriksson, S.; Tornqvist, M. Analysis of acrylamide, a carcinogen formed in heated foodstuffs. *J. Agric. Food Chem.* **2002**, *50*, 4998–5006. [CrossRef]

2. Bergmark, E. Hemoglobin Adducts of Acrylamide and Acrylonitrile in Laboratory Workers, Smokers and Nonsmokers. *Chem. Res. Toxicol.* **1997**, *10*, 78–84. [CrossRef]

3. Sorgel, F.; Weissenbacher, R.; Kinzig-Schippers, M.; Hofmann, A.; Illauer, M.; Skott, A.; Landersdorfer, C. Acrylamide: Increased concentrations in homemade food and first evidence of its variable absorption from food, variable metabolism and placental and breast milk transfer in humans. *Chemotherapy* **2002**, *48*, 267–274. [CrossRef]

4. Friedman, M. Chemistry, biochemistry, and safety of acrylamide. A review. *J. Agric. Food Chem.* **2003**, *51*, 4504–4526. [CrossRef]

5. EFSA Panel on Contaminants in the Food Chain. Scientific Opinion on acrylamide in food. *EFSA J.* **2015**, *13*, 4104. [CrossRef]

6. WHO. *IARC Monographs on the Evaluation of Carcinogenic Risks to Humans*; W. Press: Beijing, China, 1994.

7. European Commission. Council Directive 98/83/EC of 3 November 1998 on the quality of water intended for human consumption. *Off. J. Eur. Communities* **1998**, *41*, 32–54.

8. European Union. Commission Recommendation of 8 November 2013 on investigations into the levels of acrylamide in food. *Off. J. Eur. Union* **2013**, *56*, 17.

9. Yaylayan, V.A.; Wnorowski, A.; Perez Locas, C. Why Asparagine Needs Carbohydrates To Generate Acrylamide. *J. Agric. Food Chem.* **2003**, *51*, 1753–1757. [CrossRef]

10. FDA. *Guidance for Industry: Acrylamide in Foods*; Food and Drug Administration: Rockville, MD, USA, 2016.

11. Halford, N.G.; Curtis, T.Y. Reducing the Acrylamide-Forming Potential of Wheat, Rye and Potato: A Review. In *Browned Flavors: Analysis, Formation, and Physiology*; American Chemical Society: Washington, DC, USA, 2016; Volume 1237, pp. 35–53.

12. Postles, J.; Powers, S.J.; Elmore, J.S.; Mottram, D.S.; Halford, N.G. Effects of variety and nutrient availability on the acrylamide-forming potential of rye grain. *J. Cereal Sci.* **2013**, *57*, 463–470. [CrossRef]

13. Przygodzka, M.; Piskula, M.K.; Kukurová, K.; Ciesarová, Z.; Bednarikova, A.; Zieliński, H. Factors influencing acrylamide formation in rye, wheat and spelt breads. *J. Cereal Sci.* **2015**, *65*, 96–102. [CrossRef]

14. Xu, Y.; Cui, B.; Ran, R.; Liu, Y.; Chen, H.; Kai, G.; Shi, J. Risk assessment, formation, and mitigation of dietary acrylamide: Current status and future prospects. *Food Chem. Toxicol.* **2014**, *69*, 1–12. [CrossRef] [PubMed]

15. Pittet, A.; Perisset, A.; Oberson, J.M. Trace level determination of acrylamide in cereal-based foods by gas chromatography-mass spectrometry. *J. Chromatogr.* **2004**, *1035*, 123–130. [CrossRef] [PubMed]

16. Mizukami, Y.; Kohata, K.; Yamaguchi, Y.; Hayashi, N.; Sawai, Y.; Chuda, Y.; Ono, H.; Yada, H.; Yoshida, M. Analysis of acrylamide in green tea by gas chromatography-mass spectrometry. *J. Agric. Food Chem.* **2006**, *54*, 7370–7377. [CrossRef] [PubMed]

17. Soares, C.M.D.; Fernandes, J.O. MSPD Method to Determine Acrylamide in Food. *Food Anal. Methods* **2008**, *2*, 197. [CrossRef]

18. Cengiz, M.F.; Boyacı Gündüz, C.P. An eco-friendly, quick and cost-effective method for the quantification of acrylamide in cereal-based baby foods. *J. Sci. Food Agric.* **2014**, *94*, 2534–2540. [CrossRef]

19. Delevic, V.; Zejnilovic, R.; Jancic Stojanovic, B.; Djordjevic, B.; Tokic, Z.; Zrnic-Ciric, M.; Stankovic, I. Quantification of acrylamide in foods selected by using gas chromatography tandem mass spectrometry. *Hem. Ind.* **2015**, *70*, 209–215. [CrossRef]

20. Sobhi, H.R.; Ghambarian, M.; Behbahani, M.; Esrafili, A. Application of modified hollow fiber liquid phase microextraction in conjunction with chromatography-electron capture detection for quantification of acrylamide in waste water samples at ultra-trace levels. *J. Chromatogr.* **2017**, *1487*, 30–35. [CrossRef]

21. Yamazaki, K.; Isagawa, S.; Kibune, N.; Urushiyama, T. A method for the determination of acrylamide in a broad variety of processed foods by GC-MS using xanthydrol derivatization. *Food Addit. Contam.* **2011**, *29*, 705–715. [CrossRef]

22. Molina-Garcia, L.; Santos, C.S.P.; Melo, A.; Fernandes, J.O.; Cunha, S.C.; Casal, S. Acrylamide in Chips and French Fries: A Novel and Simple Method Using Xanthydrol for Its GC-MS Determination. *Food Anal. Methods* **2015**, *8*, 1436–1445. [CrossRef]

23. Luo, L.; Ren, Y.; Liu, J.; Wen, X. Investigation of a rapid and sensitive non-aqueous reaction system for the determination of acrylamide in processed foods by gas chromatography-mass spectrometry. *Anal. Methods* **2016**, *8*, 5970–5977. [CrossRef]

24. Norouzi, P.; Larijani, B.; Bidhendi, M.E.; Eshraghi, M.; Ebrahimi, M. A Sensitive Biosensor for Acrylamide Detection based on Polyaniline and Au Nanoparticles using FFT Admittance Voltammetry. *Anal. Bioanal. Electrochem.* **2018**, *10*, 18–32.

25. Zokaei, M.; Abedi, A.S.; Kamankesh, M.; Shojaee-Aliababadi, S.; Mohammadi, A. Ultrasonic-assisted extraction and dispersive liquid-liquid microextraction combined with gas chromatography-mass spectrometry as an efficient and sensitive method for determining of acrylamide in potato chips samples. *Food Chem.* **2017**, *234*, 55–61. [CrossRef] [PubMed]

26. Lagalante, A.F.; Felter, M.A. Silylation of acrylamide for analysis by solid-phase microextraction/gas chromatography/ion-trap mass spectrometry. *J. Agric. Food Chem.* **2004**, *52*, 3744–3748. [CrossRef] [PubMed]

27. Surma, M.; Sadowska-Rociek, A.; Cieślik, E. Development of a sample preparation method for acrylamide determination in cocoa via silylation. *Anal. Methods* **2016**, *8*, 5874–5880. [CrossRef]

28. Bermudo, E.; Moyano, E.; Puignou, L.; Galceran, M.T. Liquid chromatography coupled to tandem mass spectrometry for the analysis of acrylamide in typical Spanish products. *Talanta* **2008**, *76*, 389–394. [CrossRef]

29. Claus, A.; Weisz, G.M.; Kammerer, D.R.; Carle, R.; Schieber, A. A method for the determination of acrylamide in bakery products using ion trap LC-ESI-MS/MS. *Mol. Nutr. Food Res.* **2005**, *49*, 918–925. [CrossRef]

30. Tsutsumiuchi, K.; Hibino, M.; Kambe, M.; Oishi, K.; Okada, M.; Miwa, J.; Taniguchi, H. Application of ion-trap LC/MS/MS for determination of acrylamide in processed foods. *Shokuhin Eiseigaku Zasshi* **2004**, *45*, 95–99. [CrossRef]

31. Cajka, T.; Hajšlová, J. Gas Chromatography-Time-of-Flight Mass Spectrometry in Food Analysis. *LC GC Eur.* **2007**, *20*, 25.

32. Huang, Y.S.; Hsieh, T.J.; Lu, C.Y. Simple analytical strategy for MALDI-TOF-MS and nanoUPLC-MS/MS: Quantitating curcumin in food condiments and dietary supplements and screening of acrylamide-induced ROS protein indicators reduced by curcumin. *Food Chem.* **2015**, *174*, 571–576. [CrossRef]

33. Qi, Y.; Zhang, H.; Wu, G.; Zhang, H.; Gu, L.; Wang, L.; Qian, H.; Qi, X. Mitigation effects of proanthocyanidins with different structures on acrylamide formation in chemical and fried potato crisp models. *Food Chem.* **2018**, *250*, 98–104. [CrossRef]

34. Omar, M.M.; Elbashir, A.A.; Schmitz, O.J. Determination of acrylamide in Sudanese food by high performance liquid chromatography coupled with LTQ Orbitrap mass spectrometry. *Food Chem.* **2015**, *176*, 342–349. [CrossRef] [PubMed]

35. Troise, A.D.; Fiore, A.; Fogliano, V. Quantitation of Acrylamide in Foods by High-Resolution Mass Spectrometry. *J. Agric. Food Chem.* **2014**, *62*, 74–79. [CrossRef] [PubMed]

36. Pugajeva, I.; Jaunbergs, J.; Bartkevics, V. Development of a sensitive method for the determination of acrylamide in coffee using high-performance liquid chromatography coupled to a hybrid quadrupole Orbitrap mass spectrometer. *Food Addit. Contam.* **2015**, *32*, 170–179. [CrossRef] [PubMed]

37. Howell, D.C. *Fundamental Statistics for the Behavioural Sciences*, 4th ed.; Duxbury Resource Center: Pacific Grove, CA, USA, 1999.

38. European Union. European Union. European Directorate for the Quality of Medicines. In *European Pharmacopoeia*; Council of Europe: Strasbourg, France, 2007.

39. ICH, Q.R. Validation of Analytical Procedures: Text and Methodology. In Proceedings of the International Conference on Harmonization, Geneva, Switzerland, 8–13 November 2005.

40. European Commission. Commission Regulation (EU) 2017/2158 Establishing mitigation measures and benchmark levels for the reduction of the presence of acrylamide in food. *Off. J. Eur. Union* **2017**, *304*, 24–44.

41. Bratinova, S.; Karasek, L. Determination of the acrylamide content in potato chips. Report on proficiency test organised by the EURL-PAH. In *JRC Technical Reports*; European Union: Luxembourg, 2017. [CrossRef]

42. Kaufmann, A. Analytical performance of the various acquisition modes in Orbitrap MS and MS/MS. *J. Mass Spectrom* **2018**, *53*, 725–738. [CrossRef]

43. Kumar, P.; Rubies, A.; Centrich, F.; Granados, M.; Cortes-Francisco, N.; Caixach, J.; Companyo, R. Targeted analysis with benchtop quadrupole-orbitrap hybrid mass spectrometer: Application to determination of synthetic hormones in animal urine. *Anal. Chim. Acta* **2013**, *780*, 65–73. [CrossRef]

44. Pedreschi, F.; Kaack, K.; Granby, K. The effect of asparaginase on acrylamide formation in French fries. *Food Chem.* **2008**, *109*, 386–392. [CrossRef]

45. Pedreschi, F.; Mariotti, S.; Granby, K.; Risum, J. Acrylamide reduction in potato chips by using commercial asparaginase in combination with conventional blanching. *LWT-Food Sci. Technol.* **2011**, *44*, 1473–1476. [CrossRef]

46. Mustatea, G.; Popa, M.; Mioara, N. Influence of Flour Extraction Degree on Acrylamide Formation in Biscuits. *Rom. Biotechnol. Lett.* **2016**, *21*, 11328–11336.

47. Mustatea, G.; Popa, M.; Mioara, N.; Israel, F. Asparagine and sweeteners—How they influence acrylamide formation in wheat flour biscuits? *J. Biotechnol.* **2015**, *208*, S81. [CrossRef]

48. Delatour, T.; Périsset, A.; Goldmann, T.; Riediker, S.; Stadler, R.H. Improved Sample Preparation to Determine Acrylamide in Difficult Matrixes Such as Chocolate Powder, Cocoa, and Coffee by Liquid Chromatography Tandem Mass Spectroscopy. *J. Agric. Food Chem.* **2004**, *52*, 4625–4631. [CrossRef] [PubMed]

49. Van Der Fels-Klerx, H.J. Acrylamide and 5-hydroxymethylfurfural formation during baking of biscuits: NaCl and temperature–time profile effects and kinetics. *Food Res. Int.* **2014**, *57*, 210–217. [CrossRef]

50. Nguyen, H.T.; van der Fels-Klerx, H.J.I.; van Boekel, M. Acrylamide and 5-hydroxymethylfurfural formation during biscuit baking. Part II: Effect of the ratio of reducing sugars and asparagine. *Food Chem.* **2017**, *230*, 14–23. [CrossRef] [PubMed]

Foods **2019**, *8*, 597

51. Jozinović, A.; Šarkanj, B.; Ačkar, Đ.; Panak Balentić, J.; Šubarić, D.; Cvetković, T.; Ranilović, J.; Guberac, S.; Babić, J. Simultaneous Determination of Acrylamide and Hydroxymethylfurfural in Extruded Products by LC-MS/MS Method. *Molecules* **2019**, *24*, 1971. [CrossRef] [PubMed]
52. Constantin, O.E.; Kukurova, K.; Dasko, L.; Stanciuc, N.; Ciesarova, Z.; Croitoru, C.; Rapeanu, G. Modelling Contaminant Formation during Thermal Processing of Sea Buckthorn Puree. *Molecules* **2019**, *24*, 1571. [CrossRef]

 *foods*

Article

Potato Tuber Chemical Properties in Storage as Affected by Cultivar and Nitrogen Rate: Implications for Acrylamide Formation

Na Sun [1,†], Yi Wang [2], Sanjay K. Gupta [1] and Carl J. Rosen [1,*]

1   Department of Soil, Water, and Climate, University of Minnesota, St. Paul, MN 55108, USA; sunxx891@umn.edu (N.S.); gupta020@umn.edu (S.K.G.)
2   Department of Horticulture, University of Wisconsin, Madison, WI 53706, USA; wang52@wisc.edu
*   Correspondence: rosen006@umn.edu; Tel.: +1-612-625-8114
†   Na Sun, current address: Beijing Academy of Agriculture and Forestry Science, Institute of Plant Nutrition and Resources, Beijing 100097, China.

Received: 17 February 2020; Accepted: 10 March 2020; Published: 18 March 2020

Abstract: Recently released potato cultivars Dakota Russet and Easton were bred for low reducing sugars, and low acrylamide-forming potential in French fries. The objectives of this study were to determine: (1) the effects of nitrogen rate and storage time on tuber glucose concentrations in different cultivars; (2) the relationships between acrylamide, glucose, and asparagine for the new cultivars and Russet Burbank. The study was conducted at Becker, Minnesota over a period of two years on a loamy sand soil under irrigated conditions. All cultivars were subjected to five N rates from 135 to 404 kg ha^{-1} in a randomized complete block design. Following harvest, tubers were stored at 7.8 °C and sampled at 0, 16, and 32 weeks. Dakota Russet and Easton had significantly lower concentrations of stem- and bud-end glucose, asparagine, and acrylamide than those of Russet Burbank in both years. The effect of storage time on glucose concentration was significant but differed with cultivar and year. N rate effects on stem- and bud-end glucose concentrations were cultivar and storage time dependent. After 16 weeks of storage, both asparagine and acrylamide concentrations linearly increased with increasing N rate. Glucose concentration was positively correlated with acrylamide concentration ($r^2 = 0.61$). Asparagine concentration was also positively correlated with acrylamide concentration ($r^2 = 0.45$) when the asparagine:glucose ratio was <1.306. The correlation between fry color and stem-end glucose concentration was significant over three cultivars in both years, but stronger in a growing season with minimal environmental stress. Taken together, these results suggest that while acrylamide formation during potato processing is a complex process affected by agronomic practices, environmental conditions during the growing season, and storage conditions, cultivar selection may be the most reliable method to minimize acrylamide in fried products.

Keywords: reducing sugars; glucose; asparagine; acrylamide; potato; cultivar; storage time

---

1. Introduction

Acrylamide, a neurotoxin and probable carcinogen for humans, was first reported in fried potato (*Solanum tuberosum* L.) products in 2002 [1]. The concentration of acrylamide in processed products is strongly affected by processing conditions (for instance frying temperature and duration) and the concentrations of acrylamide precursors, reducing sugars, and asparagine [2,3]. Other factors such as cultivar, soil nutrition, environmental conditions during plant growth, harvesting time, storage conditions and genetic modification can affect the concentrations of reducing sugars and asparagine, and consequently acrylamide-forming potential [4–13]. Following the evaluation of numerous approaches to mitigate acrylamide in fried potato products, progress has been made in lowering its concentration. Power et al. [14] reported the mean acrylamide level decreased from 763 μg kg^{-1} in 2002 to 358 μg kg^{-1} in potato chips from 20 European countries; Wang et al. [15] showed that many new elite U.S. fry processing cultivars exist with substantially lower acrylamide-forming potential than the standard check Russet Burbank. However, despite the current mitigation efforts, acrylamide remains a public concern due to the potential cancer risk [16].

Nitrogen (N) management is a common agronomic practice that can influence acrylamide precursors, reducing sugars, glucose and fructose, and asparagine in potato tubers, and consequently acrylamide-forming potential. The effects of N rate on reducing sugars in potato tubers can be highly variable. For example, reducing sugar concentrations have been reported to both increase or decrease with increasing N supply in one study, while in another study N rate had no effect on reducing sugars [5,10]. A few studies also reported a decrease in reducing sugars at the stem end or in the whole tuber with increasing N rate [6,17–19]. Compared to the complex responses of reducing sugars to N rate, asparagine concentrations have been shown to generally increase with increasing N rate [19–21]. Low tuber reducing sugars concentrations were detected along with high asparagine concentrations in these studies [20,21].

Tuber glucose concentrations can vary significantly during storage and often interact with cultivar and growing conditions [22–25]. Muttucumaru et al. [24] reported reducing sugars' concentrations increased from 2 to 6 months of storage at 8 °C for tubers harvested from one location but decreased for tubers harvested from the other location under the same storage conditions for the cultivars Pentland Dell and Umatilla Russet. In another study, tuber glucose concentrations increased during the 9-month storage in one year but were not affected by storage time in the following year at a storage temperature of 7.2 °C for Alpine Russet [25]. The results of these studies suggest that environmental conditions during the growing season can alter or even reverse the storage time effect on reducing sugars under appropriate storage temperature conditions for certain cultivars.

Unlike the variable responses of reducing sugars to storage time, asparagine concentration is generally considered stable during storage [26–28]. Olsson et al. [29] investigated the asparagine content fluctuation during the long-term storage at both 3 and 10 °C for eight potato cultivars. Minimal effects of storage time and storage temperature on asparagine concentration were reported in the study and genetic and year effects were substantial for some cultivars. Matsuura-Endo et al. [30] reported minor variation of tuber asparagine concentrations during 18 weeks of storage at 2–18 °C.

The relationship between reducing sugars, asparagine, and acrylamide is complicated and appears to be affected by the relative concentrations of the precursors [31]. Reducing sugar concentrations are generally considered the limiting factor for acrylamide formation due to their lower concentration than asparagine in fry-processing potato tubers [26]. This relationship often results in a significantly positive correlation between reducing sugar and acrylamide concentrations [15,23,27,32,33]. However, positive correlations between asparagine and acrylamide concentrations have also been reported [30,34]. Shepherd et al. [35] suggested that both reducing sugars and asparagine should be considered to help explain the variation in acrylamide concentrations. Muttucumaru et al. [36] speculated that higher amounts of reducing sugars relative to asparagine in some cultivars were the reason for a more significant role of asparagine in acrylamide formation. They found that asparagine affected acrylamide formation when its concentration was 2.257× lower than that of reducing sugars in 20 potato cultivars grown in two locations after 2 and 6 months of storage [24].

This study investigates the effects of N fertilization rate on asparagine and glucose concentrations in recently released potato cultivars during storage and implications for acrylamide formation during processing over two growing seasons. The specific objectives of this study were to: (1) determine the effects of N rate and storage time on stem- and bud-end glucose concentrations of Easton and Dakota Russet cultivars, relative to the standard cultivar Russet Burbank over two growing seasons; and (2) elucidate relationships between acrylamide, glucose, and asparagine, for the test cultivars after 16 weeks of storage at 7.8 °C.

2. Materials and Methods

The study was conducted at the Sand Plain Research Farm in Becker, Minnesota on a Hubbard loamy sand soil (sandy, mixed, frigid Entic Hapludolls) in 2014 and 2015. A randomized complete block experimental design was adopted with a factorial treatment arrangement of N rate and cultivar replicated four times. Three French fry cultivars—Russet Burbank, Dakota Russet, and Easton—were subjected to five N fertilizer treatments, 135, 202, 269, 336, and 404 kg ha^{-1}. For each treatment, 101 kg ha^{-1} N fertilizer was applied pre-planting as polymer coated urea (44–0–0; Environmentally Smart Nitrogen, Agrium, Inc.) and 34 kg ha^{-1} N fertilizer was applied at planting as diammonium phosphate (18–46–0). The remainder, 0, 67, 134, 201, and 269 kg N ha^{-1}, for each treatment was applied at emergence as polymer coated urea.

Each plot consisted of seven 7.6 m rows with 25 plants in each row. The spacing between rows was 0.9 m and seed tubers were spaced 0.3 m apart within each row. Sample tubers were harvested from rows 4 and 5. Whole "B" seed (56–84 g) of Russet Burbank and cut "A" seed (56–84 g) of Dakota Russet and Easton were hand planted in furrows on 6 May, 2014 and 21 April, 2015. Soil properties and further cultural practices used in this study can be found in [37].

2.1. Sample Collection and Analysis

Tuber harvest dates were scheduled on 2 October 2014 and 28 September 2015 according to weather conditions. After harvest, approximately 23 kg of tubers (single tuber weight between 170 and 283 g) from each plot were shipped to the USDA-ARS (United States Department of Agriculture-Agricultural Research Service) Potato Research Worksite in East Grand Forks, Minnesota. Tubers were preconditioned at 10 °C for two weeks and then stored at 7.8 °C for 32 weeks. Glucose concentrations in both the stem and bud end of tubers were determined by a YSI-2700 Select Biochemistry Analyzer (Yellow Springs Instrument Co. Inc. Yellow Springs, Ohio, USA) after 0, 16, and 32 weeks of storage. Glucose extraction from tubers was as follow: stem-end samples (50 g per sample) were collected from the 3.8 cm of tuber tissues surrounding the stem scar, while bud-end samples (50 g per sample) were collected from the remainder of the tuber. The stem- and bud-end samples were ground separately in an Acme Juicerator (Acme Equipment, Spring Hill, FL, USA), and brought up to a final volume of 100 mL with 50 mM phosphate buffer (pH 7.2) in a beaker. Samples were then stored at 4 °C for 20–30 min and gently stirred without disturbing the precipitate (including starch) in the bottom of the beaker. For each sample, 15 mL of juice was transferred into a labeled scintillation vial and frozen for glucose analysis.

Tuber samples for asparagine analysis were collected after vine kill both years. Six tubers greater than 85 g from each plot were randomly chosen for the determination of asparagine concentration. Fresh tuber tissue was collected about 0.5 cm from the stem and bud ends of the tubers using a 7.8 mm Humboldt Brass Cork borer, and then stored at −20 °C for later asparagine extraction. Asparagine was determined by liquid chromatography with tandem mass spectrometry. Details on sample preparation and determination of asparagine analysis in this study can be found in Sun et al. [37]. All mass spectrometry analyses were conducted by the Center for Mass Spectrometry and Proteomics at the University of Minnesota.

To investigate the relationships between acrylamide formation and its precursors, we selected the storage midpoint of 16 weeks, with the assumption that asparagine concentrations analyzed at harvest were similar to those over the entire storage period [29,30]. After 16 weeks of storage, tubers were fried at the worksite in East Grand Forks. Five tubers from each plot were washed, cut (cross section dimension: 22 × 6.5 mm), blanched at 74 °C for 6 min, and then par fried in canola oil at 185 °C for 2.75 min. Even though a frying temperature of 168 °C has been recommended to lower acrylamide formation in processed potato products, our objective was to access the contribution of the precursors, asparagine and glucose, on acrylamide formation in new potato cultivars under a commercial setting. Therefore, we adopted a French fry production procedure currently used by the industry. In commercial French fry production, blanching is performed to remove precursors in the uncooked cut potatoes, which will then reduce acrylamide formation during the frying process. After freezing at −26 °C, all fries were finish fried for 1 min at 191 °C. Fry color at stem and bud ends was determined at the East Grand Forks worksite with a Photovolt Reflectance Meter (Photovolt Instruments Inc., Minneapolis, MN, USA) about 3 min after finish frying. The fried samples were shipped frozen to the University of Minnesota for acrylamide extraction. For each plot, three fries were ground for 30 s in a coffee grinder, and 0.8–1.0 g of ground powder was taken for each sample for acrylamide extraction and determination as described by Sun et al. [25]. Acrylamide was determined in whole fries and, therefore, the average concentrations of stem- and bud-end glucose and asparagine were used to explore the relationships between these precursors and acrylamide formation.

2.2. Statistical Analysis

Analysis of variance (ANOVA) for glucose in the stem and bud end as a function of N rate, cultivar, storage time, and year was conducted using PROC ANOVA (Analysis of Variance Procedure) with repeated measures for storage time in SAS 9.4 statistical software package (Statistical Analysis System, SAS Institute Inc., Cary, NC, USA). A square root transformation was used when necessary to account for the heterogeneity of variance. The average of stem-end and bud-end glucose or asparagine concentrations was considered as the concentration of the whole-tuber glucose or asparagine, which was then analyzed with PROC ANOVA as a function of N rate, cultivar, and year. Acrylamide concentrations in fried potatoes were also analyzed using PROC ANOVA. Means of interest were compared using the least significant difference (LSD) test at the 5% probability level. PROC GLM (General Linear Models Procedure) and CONTRAST statements were used to determine linear or quadratic effects of N rate on glucose, asparagine, and acrylamide concentrations. All figures and tables were depicted using Excel (Microsoft, Seattle, WA). A *p*-value < 0.05 was considered significant. The relationship between acrylamide and the molar ratio of asparagine to glucose was investigated using a piecewise linear regression (also known as a broken-stick regression) model:

$$\text{acrylamide} = a_1 + b_1 \times \text{asparagine/glucose for asparagine/glucose} \leq c \quad (1)$$

$$\text{acrylamide} = a_2 + b_2 \times \text{asparagine/glucose for asparagine/glucose} > c \quad (2)$$

where a_1 and a_2 are the intercept, b_1 and b_2 are the slope of the two linear lines, and c is the breakpoint in the model [24,38]. PROC NLIN (Nonlinear Regression Models Procedure) statement was used to estimate the parameters a_1, a_2, b_1, b_2, and c in SAS.

3. Results and Discussion

3.1. Glucose Concentrations

Stem-end glucose concentrations in 2014 and 2015 were significantly influenced by the interaction of cultivar by storage time by year (Table 1). A decrease in glucose concentration at the stem end was observed for all three cultivars during the 32-week storage in 2014 (Figure 1). The decrease of stem-end glucose concentration was significant at harvest for Easton, and 16 weeks after storage for Russet Burbank and Dakota Russet in 2014. In 2015, the fluctuation of stem-end glucose during the 32-week storage varied by cultivar (Figure 1). For Dakota Russet, the stem-end glucose concentration was stable for the first 16 weeks and significantly increased since then. However, Russet Burbank and Easton had stem-end glucose concentrations not affected by storage time in 2015. Stem-end glucose concentrations of Dakota Russet and Easton were significantly lower than that of Russet Burbank during the 32-week storage both years.

Table 1. The analysis of variance for glucose concentration during 32 weeks of storage in 2014 and 2015.

Source of Variance	Glucose During 32 Weeks of Storage	
	Stem End	Bud End
Main Effect		
Cultivar (C)	<0.0001	<0.0001
N Rate (N)	0.0718	0.4334
Year (Y)	0.1461	0.0018
Storage Time (S)	<0.0001	0.0765
Interactions		
N × Y	0.3955	0.0142
N × S	0.3319	0.7266
S × Y	<0.0001	0.0052
C × Y	<0.0001	0.0049
C × N	0.0033	0.8609
C × S	<0.0001	0.2078
N × S × Y	0.5359	0.2544
C × N × Y	0.4159	0.1309
C × N × S	0.0562	0.6043
C × S × Y	0.0295	0.2783
C × N × S × Y	0.4247	0.6849

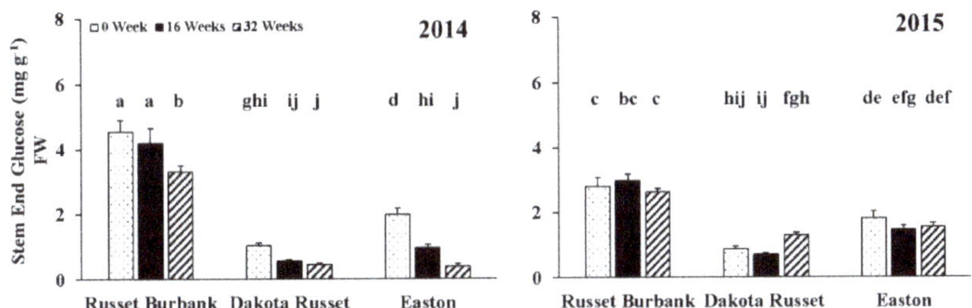

Figure 1. Three-way interaction of cultivar by storage time by year effect on stem-end glucose concentrations in 2014 and 2015 (Means were separated at the 5% level, with the same letter above the bar indicating no significant difference.). FW: Fresh weight.

The effect of N rate on stem-end glucose concentration was storage-time and cultivar dependent (Table 1 and Figure 2). Stem-end glucose concentration in Russet Burbank tended to decrease with increasing N supply through the entire storage. However, the N rate effect was significant at 32 weeks of storage only. Stem-end glucose concentrations of Easton and Dakota Russet responded to N supply at harvest only, quadratically decreased for Dakota Russet and linearly increased for Easton. Overall, the N rate effect on stem-end glucose concentration of Easton and Dakota Russet was not as dramatic as it was for Russet Burbank.

Bud-end glucose concentration of all three cultivars was significantly affected by storage time, cultivar, and N rate, but the effect differed by year (Table 1 and Figure 3). Bud-end glucose concentration averaged over cultivar and N rate decreased and then leveled off at 16 weeks of storage in 2014, but it was not affected by storage time in 2015 (Figure 3a). New cultivars Dakota Russet and Easton had significantly lower bud-end glucose concentrations than Russet Burbank in both years. However, bud-end glucose concentrations of the new cultivars significantly increased in 2015, and the differences between these two cultivars were greater with Dakota Russet than Easton in 2015 (Figure 3b). Bud-end glucose concentrations were not affected by N supply in 2014, but linearly decreased with increasing N rate in 2015 for all three cultivars averaged during the entire storage (Figure 3c). The results indicate that the effect of environmental conditions during the growing season on bud-end glucose concentration continues in storage and can be affected by N rate and cultivar.

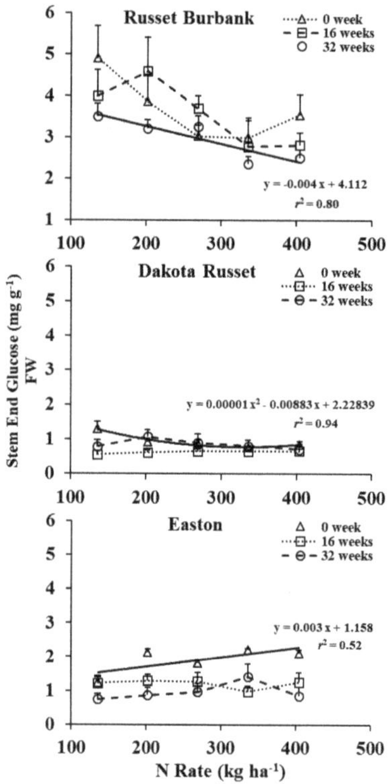

Figure 2. Three-way interaction of cultivar by N rate by storage time on stem-end glucose in 2014 and 2015.

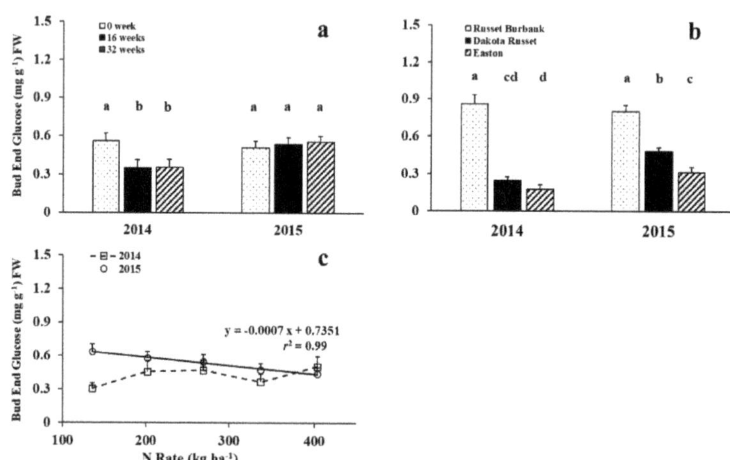

Figure 3. Interactions of storage by year (**a**), cultivar by year (**b**), and N rate by year (**c**) effects on bud-end glucose (Means were separated at the 5% level, with the same letter above the bar indicating no significant difference.).

Cold temperatures before vine kill may have induced reducing sugar accumulation in both stem and bud end for all three cultivars at harvest in 2014, while no cold stress occurred in 2015 [37]. Even though tubers were preconditioned at 10 °C for two weeks after harvest, higher glucose concentrations were still detected at the stem end for Russet Burbank in 2014 (4.54 mg g^{-1}) than in 2015 (2.79 mg g^{-1}), suggesting that two weeks was not long enough for preconditioning of this cultivar. Dakota Russet and Easton also had elevated stem-end glucose at harvest in 2014 relative to 2015 (20% higher for Dakota Russet and 9% higher for Easton), but the difference was not as large as in Russet Burbank (63%), suggesting a stress resistant characteristic of the new cultivars.

Glucose concentrations decreased in the stem and bud end of all three cultivars during storage in 2014, which may be due to the proper storage temperature (7.8 °C) for the tuber reconditioning in this study. This is consistent with the conclusion of Silva and Simon [39], who reported a decrease of glucose concentration from 2.64% to 0.58% dry weight (DW) (averaged over seven cultivars) after tuber reconditioning at 15 °C for 2 weeks during storage.

Knowles et al. [22] stored Ranger Russet, Umatilla Russet, and Russet Burbank at low temperatures (4.5 and 6.7 °C) and reported an increase in reducing sugar accumulation during the first 31 days. However, a decrease in reducing sugar concentrations was observed over the next 220 days when the storage temperature increased from 4.5 to 6.7 °C and 6.7 to 9 °C for all three cultivars. The results above suggest that sugar accumulation from cold stress is reversible with proper reconditioning temperatures.

The effect of N rate on glucose accumulation differed in stem and bud end, and often interacted with cultivar, storage time, or year. Contradictory results of N rate effects on glucose concentrations have been reported in the previous studies [10,17,40–43]. Westermann et al. [17] reported an increase in reducing sugars at the bud end, and a decrease at the stem end with increasing N supply from 0 to 336 kg ha^{-1} for Russet Burbank, which is consistent with the results for the stem end in this study. However, a minimal N rate effect on reducing sugar concentration for Russet Burbank was reported by Zebarth et al. [41]. Knowles et al. [43] reported a decreasing tendency of reducing sugars with increasing N supply at the stem and bud end for Alpine Russet, while reducing sugars of the cultivar Sage Russet barely responded to N supply at harvest. Gause [44] reported reducing sugar concentrations in Easton were not significantly affected by N rate when stored at 10 °C, which agreed in part with the results in this study. The effects of N rate on glucose concentrations in the stem and bud end of Dakota Russet have not been previously reported.

3.2. Concentrations of Acrylamide and Acrylamide Precursors

The cultivar by year interaction was significant for whole-tuber glucose concentration at 16 weeks of storage (Table 2). Due to high stem-end and low bud-end glucose concentrations in potato tuber, the whole-tuber glucose response to cultivar and year interaction was the same as that of stem-end glucose in Figure 1 (data not shown).

The main effects of cultivar, year, and N rate significantly affected whole-tuber asparagine concentrations (Table 2, Figure 4). Asparagine concentrations in Dakota Russet and Easton were consistently lower than those in Russet Burbank. Asparagine concentration increased with increasing N supply over three cultivars in two years. Averaged over cultivar and N rate, whole-tuber asparagine concentrations were lower in 2015 than in 2014, indicating the pronounced growing-condition effect. Warm weather early and late in the growing season in 2015 was favorable for tuber bulking and N uptake, but not for asparagine accumulation, suggesting that high tuber N did not proportionally convert to tuber asparagine.

Table 2. The analysis of variance for the concentrations of glucose, asparagine, and acrylamide in 2014 and 2015.

Source	Glucose (16 Weeks)	Whole-Tuber Asparagine (After Vine Kill)	Acrylamide (16 Weeks)
Main Effect			
Cultivar (C)	<0.0001	<0.0001	<0.0001
N Rate (N)	0.3042	0.0002	<0.0001
Year (Y)	0.3065	<0.0001	0.0052
Interactions			
C × N	0.1617	0.7709	0.4667
C × Y	<0.0001	0.3661	0.0049
N × Y	0.4420	0.4715	0.8615
C × N × Y	0.7170	0.9011	0.4493

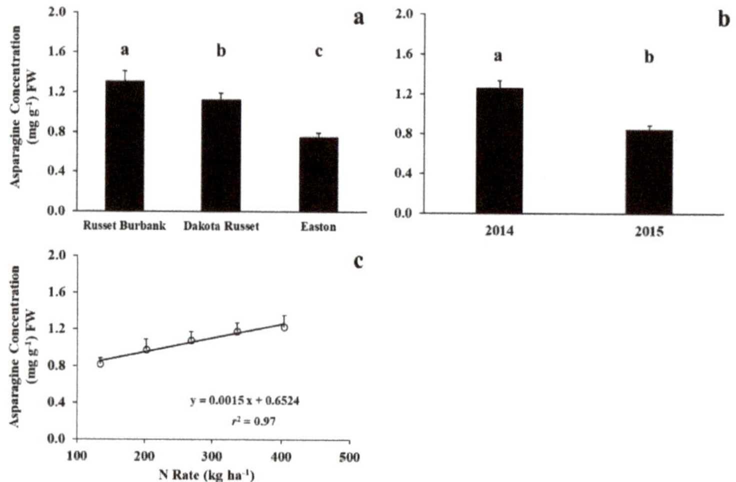

Figure 4. Main effects of cultivar (**a**), year (**b**), and N rate (**c**) on whole-tuber asparagine concentrations at 16-week storage (Means were separated at the 5% level, with the same letter above the bar indicating no significant difference.).

The effect of cultivar by year interaction was significant on acrylamide concentration after 16 weeks of storage (Table 2 and Figure 5). Russet Burbank had the same level of acrylamide concentrations in two years (388 µg kg^{-1} in 2014 and 378 µg kg^{-1} in 2015). New cultivars Dakota Russet and Easton had significantly lower acrylamide concentrations than Russet Burbank in both years. However, the acrylamide concentration was significantly higher in Dakota Russet (169 µg kg^{-1}) than in Easton (127 µg kg^{-1}) in 2014, but on an equivalent level in 2015 (209 and 203 µg kg^{-1} for Dakota Russet and Easton, respectively). The lower acrylamide concentrations in French fries produced from Dakota Russet and Easton relative to Russet Burbank are consistent with reports from the North Dakota and Maine potato breeding programs [45,46].

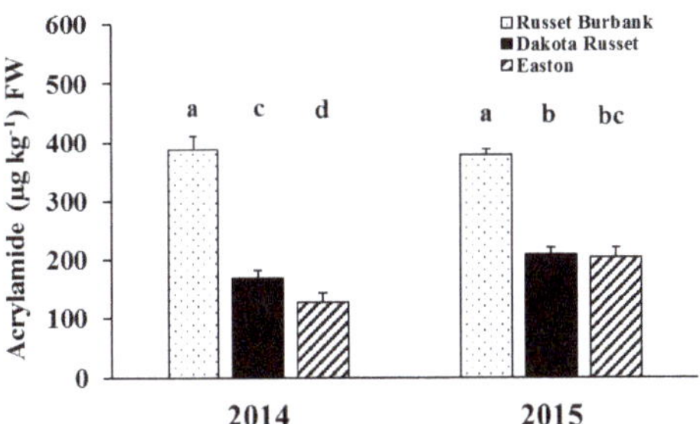

Figure 5. Cultivar by year interaction effect on acrylamide concentration at 16-week storage (Means were separated at the 5% level, with the same letter above the bar indicating no significant difference.).

Averaged over cultivar and year, acrylamide concentrations increased linearly with increasing N rate (Figure 6), which is consistent with the conclusion from previous studies on Russet Burbank by Muttucumaru et al. [10] and Easton by Gause [44]. However, in another study, acrylamide concentrations in Russet Burbank were not affected by N supply in one year, and quadratically changed in the following year [25]. These results suggest acrylamide response to N rate probably depends on environmental conditions during the growing season and cultivar. For example, acrylamide concentrations decreased in Saturna and Hermes, and increased in Lady Rosetta and Markies with increasing N supply in the United Kingdom [10] but were not affected by N supply in Switzerland for the same four cultivars [5]. The effect of N rate on acrylamide concentrations for Dakota Russet has not been previously reported.

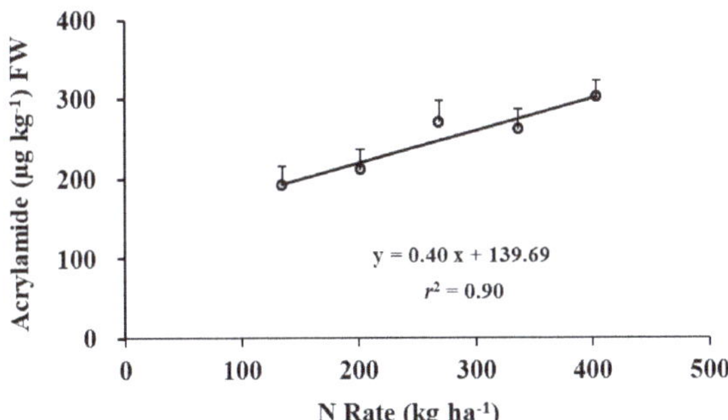

Figure 6. N rate effect on acrylamide concentration after 16 weeks of storage.

3.3. Relationships between Acrylamide, Glucose, and Asparagine

The tuber glucose concentration was significantly correlated with the concentration of acrylamide in French fries over three cultivars in two years with $r^2 = 0.61$ (Figure 7a).

The relationship between tuber asparagine and acrylamide was also significant, but not as strong ($r^2 = 0.15$, $p < 0.01$, Figure 7b). Previous studies have reported similar conclusions of a strong correlation between reducing sugars and acrylamide with R^2 values ranging from 0.73 to 0.95 [15,23,27,32,36], and a weak relationship between asparagine and acrylamide with R^2 values ranging from 0.03 to 0.27 [23,27,36].

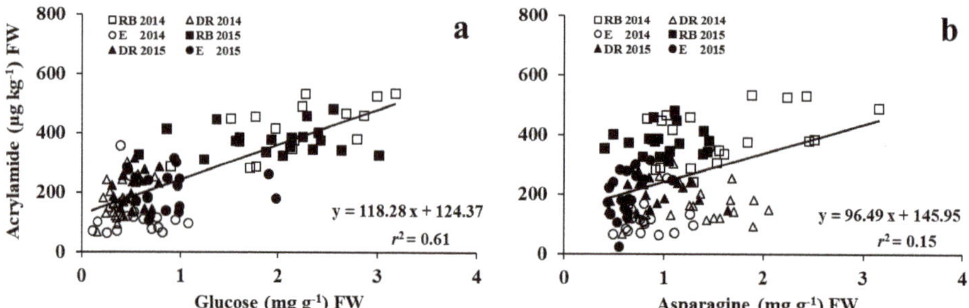

Figure 7. Relationships between acrylamide and glucose (**a**) or asparagine (**b**) concentrations of Russet Burbank, Dakota Russet, and Easton after 16 weeks of storage in 2014 and 2015.

While reducing sugar was considered the limiting factor in acrylamide formation, asparagine may play a more important role under certain circumstances [35,36]. Therefore, the relationship between acrylamide and the ratio of asparagine to glucose was analyzed using a piecewise linear regression (Figure 8). The equations for two linear lines are as follows:

$$\text{acrylamide} = 407.30 - 129.00 \times \text{asparagine/glucose for asparagine/glucose} \leq 1.306; \qquad (3)$$

$$\text{acrylamide} = 260.46 - 16.60 \times \text{asparagine/glucose for asparagine/glucose} > 1.306. \qquad (4)$$

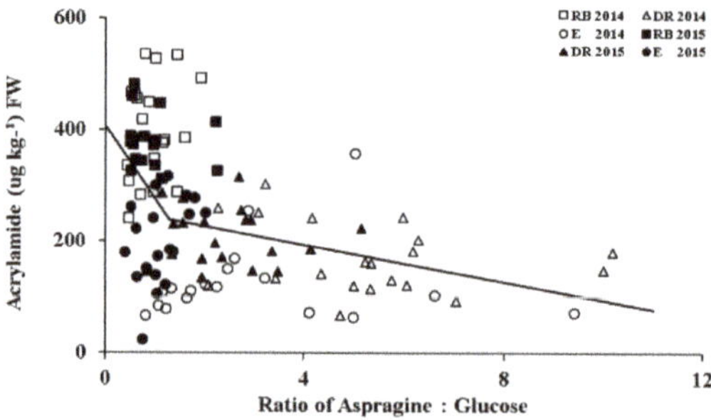

Figure 8. Relationship between acrylamide and the molar ratio of asparagine to glucose. Equations: y = 407.30 − 129.00x, x ≤ 1.306; y = 260.46 − 16.60x, x > 1.306.

When the asparagine/glucose ratio was less than 1.306, the correlation between acrylamide and asparagine was stronger, with r^2 increased from 0.15 for all data (Figure 7) to 0.45 (Figure 9). The Asparagine correlation with acrylamide was much stronger with high reducing sugar accumulation. Previous studies have also reported that the relative concentrations of reducing sugar and asparagine could affect the correlation between asparagine, reducing sugar, and acrylamide: Matsuura-Endo et al. [30] observed a stronger correlation between asparagine and acrylamide ($r^2 = 0.68$) when the molar ratio of fructose/asparagine was greater than 2. Muttucumaru et al. [24] reported a stronger correlation between asparagine and acrylamide, when concentrations of reducing sugars were 2.257-fold higher than asparagine concentrations.

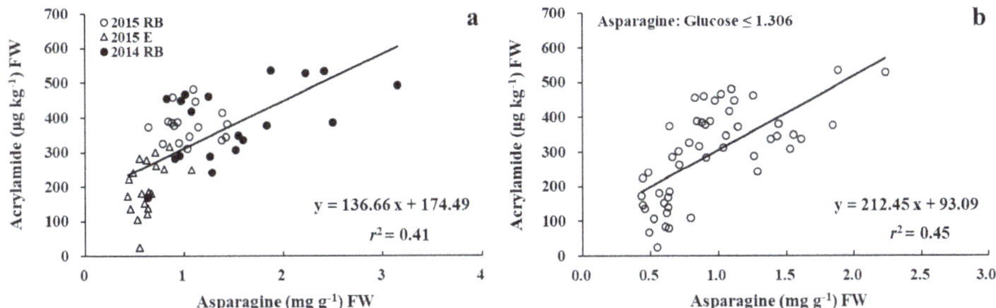

Figure 9. Relationships between acrylamide and asparagine (**a**) for Russet Burbank in two years and Easton in 2015, and (**b**) when asparagine/glucose ≤ 1.306.

In Figure 9a, most values of asparagine/glucose for Russet Burbank in two years and Easton in 2015 fell into the range of less than 1.306 as shown in Figure 9b, demonstrating a strong correlation between acrylamide and asparagine (r^2 = 0.41). The ratio of asparagine/glucose was greater than 1.306 for Dakota Russet in both years and for Easton in 2014 (data not shown). No effects of N rate on asparagine/glucose ratio were observed at 16 weeks of storage. This result suggests that cultivar and environmental conditions during the growing season were more important than N rate in affecting the relative concentrations of reducing sugars and asparagine, and consequently their relationships with acrylamide.

3.4. Relationships between Fry Color, Glucose, and Acrylamide at 16 Weeks of Storage

In the stem end, glucose concentration was significantly correlated with fry color with r^2 = 0.40 and 0.75 in 2014 and 2015, respectively (Figure 10). However, it is not the case for the bud end. The correlation between bud-end glucose and fry color was not significant in 2014 and was weak in 2015 (r^2 = 0.20). High stem-end sugar concentration generally caused dark color French fries (Figure 10a,c). Tubers containing the same amount of stem-end glucose had lower photovolt readings in 2015 than in 2014. Thus, although this correlation was significant for the stem end in both years, the equation in one year cannot be used to predict fry color or glucose concentration in another year. This is consistent with a previous study conducted with fry and chip cultivars [25].

The relationship between acrylamide and fry color was also investigated in this study. Bud-end glucose concentration was neglected due to the minimal contribution of bud-end glucose to fry color. Acrylamide and stem-end fry color was significantly correlated with an r^2 = 0.57 and $p < 0.05$ over three cultivars and two years (Figure 11), indicating that fry color can be used as a predictor of acrylamide in fried potatoes. Fry color is considered acceptable when more than 80% of French fries have photovolt readings greater than 23 [47,48]. Therefore, all tested cultivars produced French fries with acceptable color in this study.

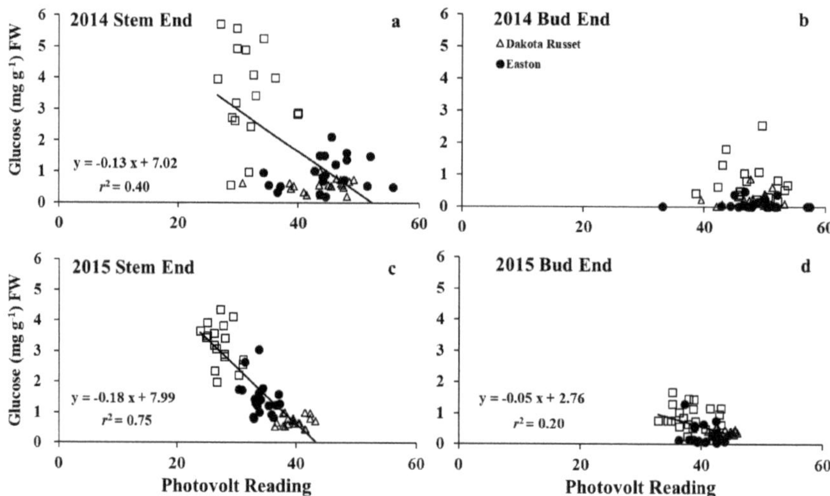

Figure 10. Relationships between tuber glucose and fry color at stem (**a,c**) and bud end (**b,d**) of French fries from tubers stored for 16 weeks at 7.8 °C in 2014 and 2015.

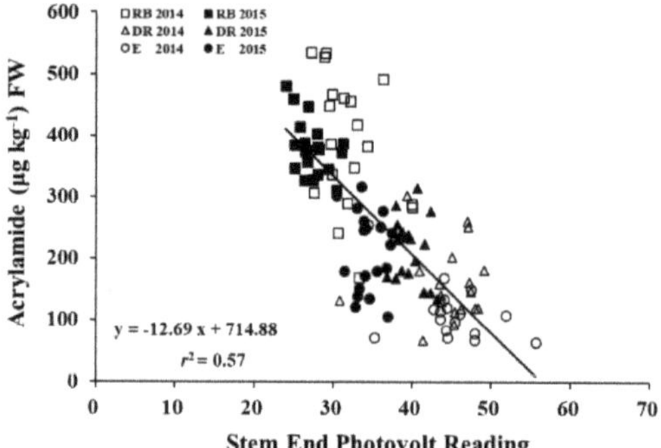

Figure 11. Relationship between stem-end fry color and acrylamide concentration of French fries from tubers stored for 16 weeks at 7.8 °C in 2014 and 2015.

4. Conclusions

New cultivars Dakota Russet and Easton contained significantly lower concentrations of acrylamide precursors (glucose at both ends and asparagine), and consequently acrylamide concentrations as well. These results suggest that cultivar selection may be the most important consideration for minimizing acrylamide during potato processing. Environmental conditions during the growing season appeared to affect tuber glucose and asparagine concentrations in storage for all three cultivars. For example, glucose concentrations decreased during storage at both ends for all three cultivars in 2014 but increased or stayed the same depending on cultivar in 2015. Lower asparagine concentration in 2015 than in 2014 was observed in all three cultivars. Differences may have been due to colder conditions during harvest in 2014 than in 2015; although other stresses during the 2014 growing season cannot be ruled out.

While glucose is generally the limiting factor in acrylamide formation, asparagine could also play an important role under high reducing sugar accumulation. With asparagine/glucose ratio <1.306, asparagine was significantly correlated with the acrylamide concentration with $r^2 = 0.45$. Cultivar and environmental conditions during the growing season seemed to affect the ratio of asparagine/glucose at 16 weeks of storage, while N rate showed no impact. Russet Burbank, Dakota Russet, and Easton produced French fries with acceptable fry color under the conditions of this study. Fry color can be used as a straightforward indicator of acrylamide and glucose concentrations in the stem end for test cultivars in this study, but the exact relationship between fry color and stem-end glucose concentration was found to differ by year.

Author Contributions: Conceptualization: C.J.R., S.K.G. and Y.W.; Funding acquisition, C.J.R.; Methodology: C.J.R., N.S., Y.W. and S.K.G; Data curation and formal analysis: N.S.; Original draft preparation, N.S.; Writing, review and editing: N.S., C.J.R., S.K.G, and Y.W. All authors have read and agreed to the published version of the manuscript.

Funding: This research was supported by a USDA-NIFA (United States Department of Agriculture-National Institute of Food and Agriculture) grant (prime award 2011-51181-30629) entitled "Improved breeding and variety evaluation methods to reduce acrylamide content and increase quality in processed potato products".

Acknowledgments: The authors would like to thank Matt McNearney, Rosa Lozano, and Ron Faber for assistance with field plot maintenance; Bruce Witthuhun for assistance with asparagine and acrylamide analysis; and Martin Glynn and Darrin Haagenson for assistance with sugar analysis.

Conflicts of Interest: The authors declare no conflict of interest.

References

1. Mottram, D.S.; Wedzicha, B.L.; Dodson, A.T. Food chemistry: Acrylamide is formed in the Maillard reaction. *Nature* **2002**, *419*, 448–449. [CrossRef]

2. Cummins, E.; Butlet, F.; Brunton, N.; Gormley, R. Factors affecting acrylamide formation in processed potato products—A simulation approach. In Proceedings of the 13th World Congress of Food Science and Technology, Nantes, France, 17–21 September 2006.

3. Visvanathan, R.; Krishnakumar, T. Acrylamide in Food Products: A Review. *J. Food Process. Technol.* **2014**, *5*, 344. [CrossRef]

4. Kooman, P.L.; Haverkort, A.J. Modelling development and growth of the potato crop influenced by temperature and daylength: LINTUL-POTATO. In Proceedings of the Potato Ecology and Modelling of Crops under Conditions Limiting Growth, Wageningen, The Netherlands, 17–19 May 1994; Volume 3, pp. 41–59.

5. Amrein, T.M.; Bachmann, S.; Noti, A.; Biedermann, M.; Barbosa, M.F.; Biedermann-Brem, S.; Grob, K.; Keiser, A.; Realini, P.; Escher, F.; et al. Potential of acrylamide formation, sugars, and free asparagine in potatoes: A comparison of cultivars and farming systems. *J. Agric. Food Chem.* **2003**, *51*, 5556–5560. [CrossRef] [PubMed]

6. Kumar, D.; Singh, B.P.; Kumar, P. An overview of the factors affecting sugar content of potatoes. *Ann. Appl. Biol.* **2004**, *145*, 247–256. [CrossRef]

7. Rommens, C.M.; Yan, H.; Sword, K.; Richael, C.; Ye, J. Low-acrylamide French fries and potato chips. *Plant Biotechnol. J.* **2008**, *6*, 843–853. [CrossRef] [PubMed]

8. Ye, J.; Shakya, R.; Shrestha, P.; Rommens, C.M. Tuber-specific silencing of the acid invertase gene substantially lowers the acrylamide-forming potential of potato. *J. Agric. Food Chem.* **2010**, *58*, 12162–12167. [CrossRef]

9. Bethke, P.C.; Bussan, A.J. Acrylamide in processed potato products. *Am. J. Potato Res.* **2013**, *90*, 403–424. [CrossRef]

10. Muttucumaru, N.; Powers, S.J.; Elmore, J.S.; Mottram, D.S.; Halford, N.G. Effects of nitrogen and sulfur fertilization on free amino acids, sugars, and acrylamide-forming potential in potato. *J. Agric. Food Chem.* **2013**, *61*, 6734–6742. [CrossRef] [PubMed]

11. Muttucumaru, N.; Powers, S.J.; Elmore, J.S.; Mottram, D.S.; Halford, N.G. Effects of water availability on free amino acids, sugars, and acrylamide-forming potential in potato. *J. Agric. Food Chem.* **2015**, *63*, 2566–2575. [CrossRef]

12. Paul, V.; Ezekiel, R.; Pandey, R. Acrylamide in processed potato products: Progress made and present status. *Acta Physiol. Plant.* **2016**, *38*, 276. [CrossRef]

13. Rosen, C.; Sun, N.; Olsen, N.; Thornton, M.; Pavek, M.; Knowles, L.; Knowles, N.R. Impact of agronomic and storage practices on acrylamide in processed potatoes. *Am. J. Potato Res.* **2018**, *95*, 319–327. [CrossRef]

14. Powers, S.J.; Mottram, D.S.; Curtis, A.; Halford, N.G. Acrylamide concentrations in potato crisps in Europe from 2002 to 2011. *Food Addit. Contam. A* **2013**, *30*, 1493–1500. [CrossRef] [PubMed]

15. Wang, Y.; Bethke, P.C.; Bussan, A.J.; Glynn, M.T.; Holm, D.G.; Navarro, F.M.; Novy, R.G.; Palta, J.P.; Pavek, M.J.; Porter, G.A.; et al. Acrylamide-forming potential and agronomic properties of elite us potato germplasm from the national fry processing trial. *Crop Sci.* **2016**, *56*, 30–39. [CrossRef]

16. European Food Safety Authority. Acrylamide in Food is a Public Health Concern. Available online: http://www.efsa.europa.eu/en/press/news/150604 (accessed on 10 January 2020).

17. Westermann, D.T.; James, D.W.; Tindall, T.A.; Hurst, R.L. Nitrogen and potassium fertilization of potatoes: Sugars and starch. *Am. Potato J.* **1994**, *71*, 433–453. [CrossRef]

18. De Wilde, T.B.; De Meulenaer, B.; Mestdagh, F.; Govaert, Y.; Vandeburie, S.; Ooghe, W.; Fraselle, S.; Demeulemeester, K.; Van Peteghem, C.; Calus, A.; et al. Influence of fertilization on acrylamide formation during frying of potatoes harvested in 2003. *J. Agric. Food Chem.* **2006**, *54*, 404–408. [CrossRef]

19. Argyropoulos, D.; Psallida, C.; Varzakas, T. The effect of nitrogen fertilisation and metabolic regulators SNRK1, GCN2 on the formation of acrylamide in two potato varieties (Spunta and Lady Rosetta) fried in corn oil. *Curr. Res. Nutr. Food Sci.* **2016**, *4*, 69–73. [CrossRef]

20. Kolbe, H. Kartoffeldüngung Unter Differenzierten ökologischen Bedingungen. Ph.D. Thesis, Georg-August-Universität, Götingen, Germany, 1990.

21. Morales, F.; Capuano, E.; Fogliano, V. Mitigation strategies to reduce acrylamide formation in fried potato products. *Ann. NY Acad. Sci.* **2008**, *1126*, 89–100. [CrossRef]

22. Knowles, N.R.; Driskill, E.P.; Knowles, L.O. Sweetening responses of potato tubers of different maturity to conventional and non-conventional storage temperature regimes. *Postharvest Biol. Technol.* **2009**, *52*, 49–61. [CrossRef]

23. Elmore, J.S.; Briddon, A.; Dodson, A.T.; Muttucumaru, N.; Halford, N.G.; Mottram, D.S. Acrylamide in potato crisps prepared from 20 UK-grown varieties: Effects of variety and tuber storage time. *Food Chem.* **2015**, *182*, 1–8. [CrossRef]

24. Muttucumaru, N.; Powers, S.J.; Elmore, J.S.; Dodson, A.; Briddon, A.; Mottram, D.S.; Halford, N.G. Acrylamide-forming potential of potatoes grown at different locations, and the ratio of free asparagine to reducing sugars at which free asparagine becomes a limiting factor for acrylamide formation. *Food Chem.* **2017**, *220*, 76–86. [CrossRef]

25. Sun, N.; Rosen, C.; Thompson, A. Acrylamide formation in processed potates as affected by cultivar, nitrogen fertilization and storage time. *Am. J. Potato Res.* **2018**, *95*, 473–486. [CrossRef]

26. De Wilde, T.; De Meulenaer, B.; Mestdagh, F.; Govaert, Y.; Vandeburie, S.; Ooghe, W.; Fraselle, S.; Demeulemeester, K.; van Peteghem, C.; Calus, A.; et al. Influence of storage practices on acrylamide formation during potato frying. *J. Agric. Food Chem.* **2005**, *53*, 6550–6557. [CrossRef] [PubMed]

27. Ohara-Takada, A.; Matsuura-Endo, C.; Chuda, Y.; Ono, H.; Yada, H.; Yoshida, M.; Kobayashi, A.; Tsuda, S.; Takigawa, S.; Noda, T.; et al. Change in content of sugars and free amino acids in potato tubers under short-term storage at low temperature and the effect on acrylamide level after frying. *Biosci. Biotechnol. Biochem.* **2005**, *69*, 1232–1238. [CrossRef] [PubMed]

28. Pavek, M.J.; Knowles, N.R. WSU Potato Cultivar Yield and Postharvest Quality Evaluations for 2015. Washington State University Special Report, 114. Available online: http://potatoes.wsu.edu/wp-content/uploads/2016/01/Potato-Cultivar-Yield-and-Postharvest-Quality-Evaluations-Research-Edition-2015.pdf. (accessed on 10 January 2020).

29. Olsson, K.; Svensson, R.; Roslund, C.A. Tuber components affecting acrylamide formation and colour in fried potato: Variation by variety, year, storage temperature and storage time. *J. Sci. Food Agric.* **2004**, *84*, 447–458. [CrossRef]

30. Matsuura-Endo, C.; Ohara-Takada, A.; Chuda, Y.; Ono, H.; Yada, H.; Yoshida, M.; Kobayashi, A.; Tsuda, S.; Takigawa, S.; Noda, T.; et al. Effects of storage temperature on the contents of sugars and free amino acids in tubers from different potato cultivars and acrylamide in chips. *Biosci. Biotechnol. Biochem.* **2006**, *70*, 1173–1180. [CrossRef] [PubMed]

31. Halford, N.G.; Curtis, T.Y.; Muttucumaru, N.; Postles, J.; Elmore, J.S.; Mottram, D.S. The acrylamide problem: A plant and agronomic science issue. *J. Exp. Bot.* **2012**, *63*, 2841–2851. [CrossRef]

32. Amrein, T.M.; Schönbächler, B.; Rohner, F.; Lukac, H.; Schneider, H.; Keiser, A.; Escher, F.; Amadò, R. Potential for acrylamide formation in potatoes: Data from the 2003 harvest. *Eur. Food Res. Technol.* **2004**, *219*, 572–578. [CrossRef]

33. Halford, N.G.; Muttucumaru, N.; Powers, S.J.; Gillatt, P.N.; Hartley, L.; Elmore, J.S.; Mottram, D.S. Concentrations of free amino acids and sugars in nine potato varieties: Effects of storage and relationship with acrylamide formation. *J. Agric. Food Chem.* **2012**, *60*, 12044–12055. [CrossRef]

34. Zhu, F.; Cai, Y.Z.; Ke, J.; Corke, H. Compositions of phenolic compounds, amino acids and reducing sugars in commercial potato varieties and their effects on acrylamide formation. *J. Sci. Food Agric.* **2010**, *90*, 2254–2262. [CrossRef]

35. Shepherd, L.V.T.; Bradshaw, J.E.; Dale, M.F.B.; McNicol, J.W.; Pont, S.D.A.; Mottram, D.S.; Davies, H.V. Variation in acrylamide producing potential in potato: Segregation of the trait in a breeding population. *Food Chem.* **2010**, *123*, 568–573. [CrossRef]

36. Muttucumaru, N.; Powers, S.J.; Elmore, J.S.; Briddon, A.; Mottram, D.S.; Halford, N.G. Evidence for the complex relationship between free amino acid and sugar concentrations and acrylamide-forming potential in potato. *Ann. Appl. Biol.* **2014**, *164*, 286–300. [CrossRef] [PubMed]

37. Sun, N.; Wang, Y.; Gupta, S.K.; Rosen, C.J. Nitrogen fertility and cultivar effects on potato agronomic properties and acrylamide-forming potential. *Agron. J.* **2019**, *3*, 408–418. [CrossRef]

38. Ryan, S.E. A Tutorial on the Piecewise Regression Approach Applied to Bedload Transport Data. USDA Forest Service—General Technical Report RMRS-GTR-189. Available online: https://www.fs.usda.gov/rds/archive/Product/RDS-2007-0004/ (accessed on 10 January 2020).

39. Silva, E.M.; Simon, P.W. Genetic, physiological, and environmental factors affecting acrylamide concentration in fried potato products. In *Chemistry and Safety of Acrylamide in Food*; Friedman, M., Mottram, D., Eds.; Springer: Boston, MA, USA, 2005; Volume 561, pp. 371–386.

40. Long, C.M.; Snapp, S.S.; Douches, D.S.; Chase, R.W. Tuber yield, storability, and quality of Michigan cultivars in response to nitrogen management and seed piece spacing. *Am. J. Potato Res.* **2004**, *81*, 347–357. [CrossRef]

41. Zebarth, B.J.; Leclerc, Y.; Moreau, G.; Botha, E. Rate and timing of nitrogen fertilization of Russet Burbank potato: Yield and processing quality. *Can. J. Plant Sci.* **2004**, *84*, 855–863. [CrossRef]

42. Gerendás, J.; Heuser, F.; Sattelmacher, B. Influence of nitrogen and potassium supply on contents of acrylamide precursors in potato tubers and on acrylamide accumulation in French fries. *J. Plant Nutr.* **2007**, *30*, 1499–1516. [CrossRef]

43. Knowles, N.R.; Pavek, M.J.; Knowles, L.O. Developmental profiles, nitrogen use and postharvest quality of alpine and sage russet potatoes in the columbia basin. In Proceedings of the Annual Washington and Oregon Potato Conference, Kennwick, WA, USA, 27–30 January 2015; pp. 37–50. Available online: http://www.nwpotatoresearch.com/IPMStuff/PDFs/Proceedings2015.pdf (accessed on 10 January 2020).

44. Gause, K. Effect of Nitrogen and Potassium on Potato Yield, Quality and Acrylamide-Forming Potential. Master's Thesis, University Maine, Orono, ME, USA, 2014.

45. North Dakota State University Research Foundation. Available online: http://www.ndsuresearchfoundation.org/dakota_russet (accessed on 10 January 2020).

46. Porter, G.; Alyokhin, A.; Lambert, D.; Halseth, D.; Yencho, G.; Jemison, J.; Freeman, J.; Perry, K.; Perkins, L.B.; Zotarelli, L.; et al. *Potato Breeding and Variety Development for Improved Quality and Pest Resistance in the Eastern United States*; Grant reports; University of Maine office of research and sponsored programs: Orono, ME, USA. Available online: http://digitalcommons.library.umaine.edu/orsp_reports/31/. (accessed on 10 January 2020).

47. Shock, C.C.; Stieber, T.D.; Zalewski, J.C.; Eldredge, E.P.; Lewis, M.D. Potato tuber stem-end fry color determination. *Am. J. Potato Res.* **1994**, *71*, 77–88. [CrossRef]

48. Driskill, E.P.; Knowles, L.O.; Knowles, N.R. Temperature-induced changes in potato processing quality during storage are modulated by tuber maturity. *Am. J. Potato Res.* **2007**, *84*, 367–383. [CrossRef]

Article

Formation of Acrylamide and Other Heat-Induced Compounds during Panela Production

Marta Mesias [1], Cristina Delgado-Andrade [1,*], Faver Gómez-Narváez [2], José Contreras-Calderón [2] and Francisco J. Morales [1]

[1] Instituto de Ciencia y Tecnología de Alimentos y Nutrición (ICTAN-CSIC), C/José Antonio Novais, 10, E-28040 Madrid, Spain; mmesias@ictan.csic.es (M.M.); fjmorales@ictan.csic.es (F.J.M.)

[2] Bioali Research Group, Food Department, Faculty of Pharmaceutical and Food Sciences, University of Antioquia, Calle 67 No. 53–108, Ciudad Universitaria, Medellín 050010, Colombia; faver.gomez@udea.edu.co (F.G.-N.); jose.contrerasc@udea.edu.co (J.C.-C.)

* Correspondence: cdelgado@ictan.csic.es; Tel.: +34-915492300

Received: 6 March 2020; Accepted: 17 April 2020; Published: 22 April 2020

Abstract: Non-centrifugal cane sugar (panela) is an unrefined sugar obtained through intense dehydration of sugarcane juice. Browning, antioxidant capacity (measured by ABTS (2,2'-azino-bis (3-ethylbenzothiazoline-6-sulfonic acid) assay and total phenolic content) and the formation of acrylamide and other heat-induced compounds such as hydroxymethylfurfural (HMF) and furfural, were evaluated at different stages during the production of block panela. Values ranged between below the limit of quantitation (LOQ)–890 µg/kg, < LOQ–2.37 mg/kg, < LOQ–4.5 mg/kg, 0.51–3.6 Abs 420 nm/g, 0.89–4.18 mg gallic acid equivalents (GAE)/g and 5.08–29.70 µmol TE/g, for acrylamide, HMF, furfural, browning, total phenolic content and ABTS (all data in fresh weight), respectively. Acrylamide significantly increased as soluble solid content increased throughout the process. The critical stages for the formation of acrylamide, HMF and furfural were the concentration of the clarified juice in the concentration stage to get the panela honey and the final stage. Similar trends were observed for the other parameters. This research concludes that acrylamide, HMF and furfural form at a high rate during panela processing at the stage of juice concentration by intense evaporation. Therefore, the juice concentration stage is revealed as the critical step in the process to settle mitigation strategies.

Keywords: panela; processing; acrylamide; furanic compounds; antioxidants; non-enzymatic browning

---

1. Introduction

Production of panela, also known as unrefined non-centrifugal sugar or non-centrifugal cane sugar, is one of the most traditional agro-industries in tropical countries. Panela is obtained by grinding the sugar cane, clarifying, evaporating the juice and concentrating it until honey is obtained (more than 90° Brix). This is then beaten, molded and cooled to achieve solidification [1]. Panela production exhibits yields between 6.4% and 14.9% [2].

Panela is produced by small farmers around the world. India is by far the world's largest producer of panela, accounting for 56.15% of global production in 2011. Colombia, the world's second largest producer, contributed 14.1% of total panela globally in 2011. While India and Colombia dominate global production, other countries from Asia, Africa or Latin America are also producers of this foodstuff [3]. The use of panela in industrially produced foods is as a sweetener to replace refined sugars, but also as an ingredient for the manufacture of foodstuffs such as puddings, baked goods, marmalades, protein/energy/weight control bars, desserts, confectionary and chocolate products [4].

Unlike white sugar, panela contains minerals, vitamins, phenolic compounds, amino acids and proteins. The presence of bioactive, health-promoting compounds increases the beneficial effects of panela on human health, including effects that are anticarcinogenic, antitoxic, cytoprotective, anti-inflammatory and antiatherogenic [5].

Due to high sugar ($\approx$14 g/100 g) and nitrogen compound ($\approx$0.40 g/100 g) levels in sugarcane [6], the Maillard reaction and caramelization are the main chemical reactions to take place during panela production. The Maillard reaction is promoted at >50 °C and pH 4–7 and occurs in low moisture conditions. The initial step involves the formation of a Schiff base from the reaction of the amino group on an amino acid with the carbonyl group of a reducing sugar, which can rearrange to form an Amadori compound whose degradation yields intermediate compounds such as acrylamide and furanic compounds. Caramelization is favored at temperatures of >120 °C and pH 3–9 and involves another nonenzymatic browning step through the degradation of reducing sugars without the condensation step [7]. Acrylamide formation in panela is possible due to the presence of free amino acids and reducing sugars, intense thermal treatment, and the low moisture found in the final product [1,6]. Acrylamide has been described to present neurotoxic, genotoxic, carcinogenic and reproductive toxic effects [8,9]. On the other hand, furanic compounds such as hydroxymethylfurfural (HMF) and furfural have been extensively applied as heat-induced chemical indexes for monitoring the thermal damage of food. These furan derivatives, which can be generated during panela production through both caramelization and Maillard reactions [7,10], have been confirmed to confer genotoxic, mutagenic, carcinogenic, DNA-damaging, organotoxic and enzyme inhibitory effects. It has been reported that HMF is an indirect mutagen because it is converted to an active metabolite, the sulfuric acid ester 5-sulfo-oxymethylfurfural (SMF), with mutagenicity [11], whereas furfural has shown toxicological effects leading to hepatotoxicity [12].

Scientific research is scarce in relation to the occurrence and pathways involved in the formation of heat-induced process contaminants, particularly acrylamide and furfurals, during panela production. In addition, data have mostly been derived from methodologies that apply analytical techniques which lack sensitivity and fail to refer to normalized procedures [1]. Due to the toxicological effects on human health related to these compounds, evaluation of their formation in foods is necessary to look for mitigation strategies aimed to reduce the exposure to the contaminants. Thus, the aim of this work was to study the impact of different stages during the panela production chain on the formation of acrylamide and other heat-induced process contaminants by applying robust and validated analytical methods. The results will provide greater insight into the identification of critical points in the process. It is the first time that such assays have been carried out together and, therefore, the outcomes of this investigation could be key for establishing mitigation strategies which can be used by panela producers and food safety bodies.

2. Material and Methods

2.1. Chemicals and Reagents

Potassium hexacyanoferrate (II) trihydrate (98%, Carrez-I) and zinc acetate dehydrate (>99%, Carrez-II) were obtained from Sigma (St. Louis, MO, USA). $^{13}C_3$-labelled acrylamide (99% isotopic purity) was obtained from Cambridge Isotope Laboratories (Andover, MA, USA. Formic acid (98%), D(+) glucose, D(-) fructose, D(-) sorbitol, ethanol, methanol (99.5%) and hexane were obtained from Panreac (Barcelona, Spain). Deionized water was obtained from a Milli-Q Integral 5 water purification system (Millipore, Billerica, MA, USA). All other chemicals, solvents and reagents were of analytical grade.

2.2. Samples

Samples were supplied by a large panela producer located in the province of Antioquia (Colombia). Samples were collected at four different stages, which represent critical points during the manufacture of block panela (Figure 1). Samples were as follows: raw cane juice obtained after the cane is ground (Sample 1); clarified juice obtained by heating (<100 °C) sample 1 (Sample 2); concentrated juice produced at 60–65° Brix following evaporation (>110 °C) (Sample 3); block panela obtained following concentration (>120 °C) (Sample 4). Evaporation of the juice is performed in batches of 1000 L, whilst being heated at 110–120 °C for approximately 40 min. Panela honey concentration is performed at 120 °C for approximately 20 min in batches of 40 L. Samples 1, 2 and 3—which correspond to the juices—were lyophilized and stored at −20 °C whilst awaiting analysis. Block panela (sample 4) was stored at room temperature.

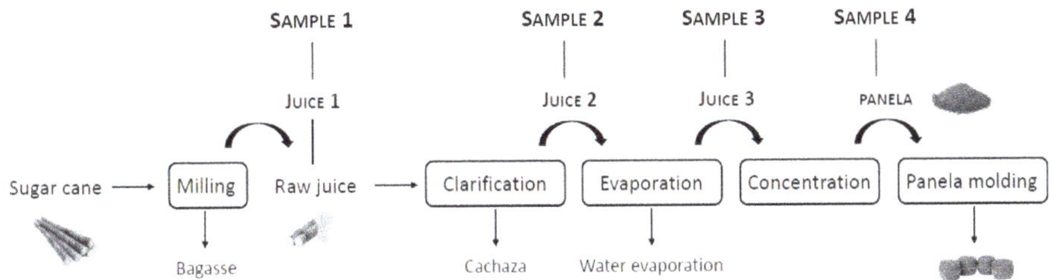

Figure 1. Scheme of the panela production process.

2.3. Basic Analyses

Moisture was estimated gravimetrically to constant weight following well-established procedures for panela juices [13] and block panela [14]. The pH of panela was determined by mixing the sample (1 g) with 100 mL of water and vortexing for 3 min. The mixture was kept at room temperature for 1 h and centrifuged to separate impurities. The pH was measured using a CG-837 pH meter (Schott, Mainz, Germany). The pH of juices was directly measured in fresh samples immediately following collection. Soluble solids content (°Brix; in original juices immediately after their collection) were measured with a digital refractometer OptiDuo (Bellingham + Stanley, Kent, UK). Soluble solid content in panela was calculated from the moisture content.

2.4. Determination of Reducing Sugars

The content of reducing sugars (glucose + fructose) was determined in lyophilized juices and in the panela sample via high performance liquid chromatography using a refractive index detector (HPLC-RID). The procedure was based on a slightly modified version of the method described by Ayvaz [15]. Three hundred mg of sample was weighed and mixed with 9 mL of 80% (*v/v*) ethanol and 1 mL of sorbitol (10 mg/mL), as an internal standard. Following vortex agitation, the mixture was incubated at 50 °C and 900 rpm for 1 h, and centrifuged at 4 °C and 5000 rpm for 20 min. The supernatant was transferred into a new tube and ethanol was evaporated using TurboVap equipment (Biotage, Uppsala, Sweden). The aqueous extract was purified via solid-phase extraction using an SCX cartridge (Supelco, Sigma Aldrich, St. Louis, MO, USA) and filtered (0.22 µm pore-size membrane) prior to HPLC analysis. Twenty µL of extract was injected into the HPLC System LC-20 AD, using a RID-10A (Shimadzu, Scientific Instruments, INC, Columbia, MD, USA). Analytical separation was achieved with a Rezex RCM-Monosaccharide Ca^{2+} column (300 × 7.8 mm, 8 µm; Phenomenex, Torrance, CA, USA) at 80 °C in isocratic elution, with a mobile phase of deionized water and a flow rate of 0.6 mL/min. Sugars were quantified using standard solutions spiked with sorbitol. Results were expressed as g/100 g of fresh weight (FW) and dry matter (DM). The analysis was performed in duplicate.

2.5. Determination of Asparagine

Asparagine was determined in the lyophilized juices and panela sample via gas chromatography-flame ionization detection (GC-FID), according to Farkas and Toulouee [16] but with some minor modifications as described by Mesias et al. [17]. A GC-FID (Agilent GC 7820A FID) equipped with an automatic injector was used for quantitation. An amino acid dedicated column (Zebron ZBAAA capillary; 10 × 0.25 mm) was used to separate amino acids. Starting oven temperature was set at 110 °C and increased 32 °C per minute until 320 °C was reached. An aliquot of the derivatized sample (1 μL) was injected in split mode (15:1) at 250 °C. The FID detector was set to 320 °C and the carrier helium gas flow rate was maintained at 1.5 mL/min whilst in process. External calibration was carried out using asparagine standard and results were corrected according to norvaline recovery, this being used as an internal standard. Free asparagine content was expressed as mg/100 g of FW and DM. Analysis was performed in duplicate.

2.6. LC-ESI-MS-MS Determination of Acrylamide

Sample extraction followed a slightly modified version of the method described by Mesias and Morales [18]. Lyophilized juice and panela samples (0.5 g) were weighed and mixed with 9.4 mL of water in polypropylene centrifugal tubes. The mixture was spiked with 100 μL of a 5 μg/mL [$^{13}C_3$]-acrylamide methanolic solution, which served as an internal standard, and later homogenized (Ultra Turrax, IKA, Mod-T10 basic, Bohn, Germany) for 10 min. Afterwards, samples were treated with 250 μL of Carrez I (15 g potassium ferrocyanide/100 mL water) and Carrez II (30 g zinc acetate/100 mL water) solutions, and centrifuged (9000 g for 10 min) at 4 °C. Samples were clarified using Oasis-HLB cartridges (Supelco, Saint Louis, MO, USA) and extracts were analyzed according to Mesías and Morales [18]. Acrylamide recovery occurred between 90% and 106%. Relative standard deviations (RSD) for precision, repeatability and reproducibility of the analyses were calculated as 2.8%, 1.2% and 2.5%, respectively. The procedure fulfilled method performance requirements established by the EU acrylamide Regulation 2017/2158. The limit of quantitation was set at 20 μg/kg. Acrylamide results were expressed as μg/kg of FW and DM, and comparisons were made between different stages. Samples were analyzed in duplicate.

2.7. Determination of HMF and Furfural

HMF and furfural content was determined in lyophilized juices and panela samples using High-Performance Liquid Chromatography with Diode-Array Detection (HPLC-DAD), as described by Mesías et al. [19]. The limits of quantification were set at 0.6 and 0.3 mg/kg for HMF and furfural, respectively. Results were expressed as mg/kg of FW and DM. Samples were analyzed in duplicate.

2.8. Determination of Browning

Supernatant fractions (200 μL) obtained during preparation of HMF and furfural samples were placed in 96-well plates. Browning at 420 nm was measured at room temperature using a BioTekSynergyTM HT-multimode microplate spectrophotometer (BioTek Instruments, Winooski, VT, USA). Samples were analyzed in duplicate and results were expressed as absorbance units (AU)/g of FW and DM.

2.9. Sample Extraction for Measurement of Antioxidant Activity (ABTS (2,2′-azino-bis(3-ethylbenzothiazoline-6-sulfonic acid) Assay and Total Phenolic Content)

Sample extraction was performed following the procedure described by Pérez-Jiménez and Saura-Calixto [20]. Briefly, 0.1 g of lyophilized juice and panela sample was placed in a tube, and 6 mL of acidic methanol/water (50:50 *v/v*, pH 2) was added. The tube was thoroughly shaken at room temperature for 20 min and centrifuged at 2500 g for 10 min in order to recover the supernatant. Four milliliters of the same acidic methanol/water solution were added to the residue, with shaking and centrifugation steps then being repeated. The second methanolic extract was combined with the first one. When necessary, proper dilutions with distilled water were performed to measure in the ABTS assay and the total phenolic content. Extraction was performed in duplicate.

2.9.1. Total Phenolic Content (TPC)

Total phenolic content was determined according to a slightly modified version of the Folin–Ciocalteu method described by Marfil et al. [21]. Briefly, 80 µL of sample, blank or gallic acid standard, 1520 µL of distilled water and 300 µL of 20% Na_2CO_3 (w/v) were mixed with 100 µL of commercial Folin–Ciocalteu's reagent and incubated for 1h at room temperature. Absorbance was measured at 750 nm using a Biotek Synergy HT multi-mode microplate reader (BioTek® Instruments Inc., USA). Results were expressed as mg gallic acid equivalent (GAE)/g of FW and DM. All measurements were performed in triplicate.

2.9.2. ABTS Assay

The ABTS assay was developed as described by Rufián-Henares and Delgado-Andrade [22], with slight modifications. The ABTS+• was produced by reacting 7 mM ABTS stock solution with 2.45 mM potassium persulfate and allowing the mixture to stand in the dark at room temperature for 12–16 h before use. The ABTS+• working solution (stable for 2 days) was diluted with an ethanol: water (50:50) solution until an absorbance of 0.70 ± 0.02 at 730 nm was achieved. For analyses, 40 µL of sample, blank or Trolox standard and 200 µL of 5 mM pH 8.4 phosphate buffer were added to 60 µL of diluted ABTS+• solution. The absorbance reading was taken at 10 min using the previously described microplate reader. Aqueous Trolox solutions were used for calibration (15–125 µM). Results were expressed as µmol Trolox equivalents (TE)/g of FW and DM. All measurements were performed in triplicate.

2.10. Statistical Analysis

One-way ANOVA was used to investigate differences between final block panela and sugar cane juice, in the content of processing contaminants and other physicochemical variables at three different processing stages. Significant differences were established using the LSD test, with a confidence of 95%. All statistical analyses were performed using Statgraphics Centurion® Version XVI (The Plains, VA, USA).

3. Results and Discussion

Samples collected at different stages during panela processing were analyzed for pH value, moisture and soluble solids content (Table 1). According to the good processing practices for high quality block panela [6], the pH of the mature sugar cane must increase from 5.0–5.3 to 5.2–5.4 after crushing (raw juice cane). Then, during the clarification stage, the pH of the juice will remain between 5.8 and 6.5 by the addition of pH regulators that will avoid sucrose inversion [6]. Similarly, the soluble solids in the raw juice must be ≥20° Brix, and between 18 and 20° Brix in the juice following clarification. After the heating and evaporation stage, the pH of the panela honey must be at least 5.8, with the soluble solids around 65–70° Brix. At the final stage, the panela normally exhibits pH values between 5.6 and 6.3, and a soluble solid content between 88° and 92° Brix. In the present study, results of pH and soluble solid content are in line with that described above for block panela. Sample 1 (raw juice) showed a pH value of 5.4, which increased to 6.0 following pH regulation. In the same way, soluble solid content increased from 21.2 in sample 1, to 92.4° Brix in the block panela. As expected, the moisture content decreased from 78.8 to 7.6% as panela elaboration progressed.

Table 1. Moisture, soluble solids, pH and acrylamide precursors (reducing sugars and asparagine) in samples obtained during panela processing (fresh weight—FW).

	Sample 1	Sample 2	Sample 3	Sample 4
Moisture (%)	78.8 [c]	81.5	32.5	7.60
Soluble solids (° Brix)	21.2	18.5	67.5	92.4
pH	5.40	5.50	5.70	6.00
Reducing sugars (g/100 g)	3.20 [b] (3.18–3.22)	2.65 [a] (2.53–2.78)	4.01 [c] (3.89–4.12)	6.02 [d] (6.02–6.03)
Asparagine (mg/100 g)	16.2 [a] (15.5–17.0)	12.7 [a] (12.4–12.9)	80.5 [c] (68.7–92.3)	39.7 [b] (38.4–41.0)

Results are expressed as mean (range) ($n = 2$). Different superscripts in the same row indicate significant differences ($p < 0.05$).

3.1. Acrylamide

The acrylamide content ranged from < limit of quantitation (LOQ) to 890 µg/kg sample (FW) (Table 2). Throughout the panela manufacturing progress, the formation of acrylamide increased from sample 1 to sample 4 (panela). This increase in acrylamide formation is in line with the use of higher temperatures and the relative concentration of solids in the samples. At the end of the concentration step (Figure 1), sample 3 (67.5° Brix) exhibited an acrylamide content of 298 µg/kg, with this climbing sharply to reach 890 µg/kg (Table 2) in the final product (92.4° Brix). In line with Vargas Lasso et al. [1], negligible amounts of acrylamide (<LOQ) were detected in raw cane juice and in the juice following clarification. Vargas Lasso et al. [1] reported higher values in concentrated juice (800 µg/kg) and final block panela (2200 µg/kg) than those observed in our study. The differences with these authors could be related to the different origins of sugar cane, amounts of acrylamide precursors and processing conditions. However, application of a non-specific and non-selective analytical technique for acrylamide should also be considered as this could lead to overestimations instead of detections based on tandem mass spectrometry. The fate of acrylamide in the samples corroborates the conclusion that temperature and moisture content are the key physical parameters to influence acrylamide formation during panela production, whilst maintaining levels of precursors in sugar cane.

Table 2. Acrylamide, hydroxymethylfurfural (HMF) and furfural content, browning (420 nm), ABTS (2,2′-azino-bis(3-ethylbenzothiazoline-6-sulfonic acid) and total phenolic content (TPC) in samples obtained during panela processing (FW).

	Sample 1	Sample 2	Sample 3	Sample 4
Acrylamide (µg/kg)	<LOQ	<LOQ	298 ± 31 [a]	890 ± 15 [b]
HMF (mg/kg)	<LOQ	<LOQ	<LOQ	2.37 ± 0.16
Furfural (mg/kg)	<LOQ	<LOQ	<LOQ	4.50 ± 0.75
Browning (AU/g)	0.51 [a] (0.50–0.53)	0.41 [a] (0.40–0.41)	1.70 [b] (1.69–1.70)	3.60 [c] (3.53–3.67)
ABTS (µmol TE/g)	5.08 ± 0.13 [b]	3.47 ± 0.15 [a]	20.68 ± 0.78 [c]	29.70 ± 0.41 [d]
TPC (mg GAE/g)	0.89 ± 0.02 [b]	0.66 ± 0.01 [a]	3.12 ± 0.04 [c]	4.18 ± 0.08 [d]

Results are expressed as mean (range) ($n = 2$) and as mean ± SD ($n = 3$). Different superscripts in the same row indicate significant differences ($p < 0.05$).

Data available in the literature are scarce regarding acrylamide content in panela. Hoenicke and Gatermann [23] found values of around 140 µg/kg in raw sugar. These data are much lower than that detected in the present study. This may be attributed to reasons such as a more drastic thermal process, high content of precursors in raw juice and inadequate pH regulation, which favors sucrose hydrolysis followed by the Maillard reaction. The acrylamide level measured in the block panela sample was close to the highest values reported by Gómez-Narváez et al. [24] in eight panela blocks, with an average figure of 540 µg/kg.

Figure 2 shows the trends in acrylamide, asparagine and reducing sugars (results expressed as DM) at critical stages of the process. No significant changes ($p > 0.05$) were observed in sample 1 and sample

2 regarding acrylamide and its precursors. However, the acrylamide content increased, whilst the content of its precursors decreased during the concentration stage. It is unexpected that asparagine content increased in sample 3 (expressed in DM), however, standard deviations between measurements were much higher when compared to the other samples. Residual proteolysis activity following the breakage of vegetable cells is also plausible during crushing and clarification stages at moderate temperatures, prior to inactivation induced by temperatures higher than 100 °C and the resulting increase in asparagine. The concentration stage during panela production is characterized by intense heat treatment during which clarified juice is boiled at temperatures higher than 120 °C. Following this, a marked drop in asparagine was displayed, together with a high increase in the acrylamide concentration in block panela. With regard to the reducing sugars content, a trend towards a decrease is shown, with a steeper decline observed between sample 2 and sample 3. The stable levels seen in reducing sugars between sample 3 and sample 4 (panela) would suggest that acrylamide formation during this stage occurs through several reaction mechanisms. These mechanisms occur in low humidity conditions and involve decarboxylation of the Schiff base, leading to Maillard intermediates that directly or indirectly release acrylamide [25]. The reducing sugar content patterns in panela are similar to those reported by other authors. Jaffé [26] described levels of between 3.69 and 10.5 g/100 g, whereas Lee et al. [27] indicated ranges of 2.10–3.30 g/100 g for glucose and 1.76–2.56 g/100 g for fructose. These high reducing sugar levels suggest that asparagine could be the limiting factor of acrylamide formation in panela samples.

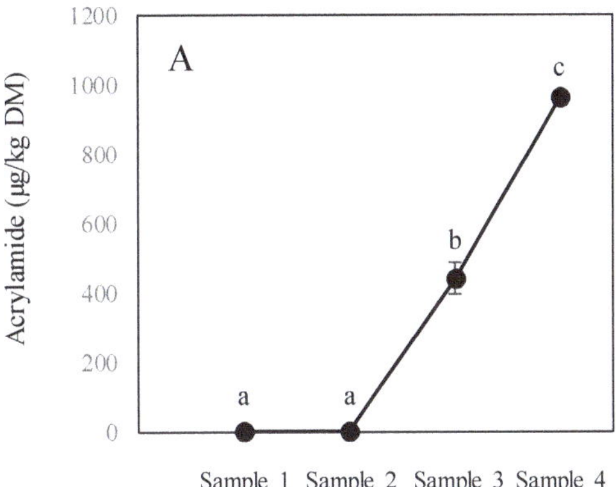

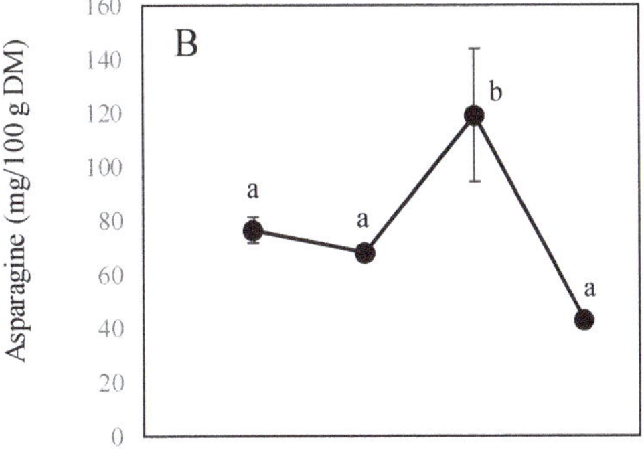

Figure 2. *Cont.*

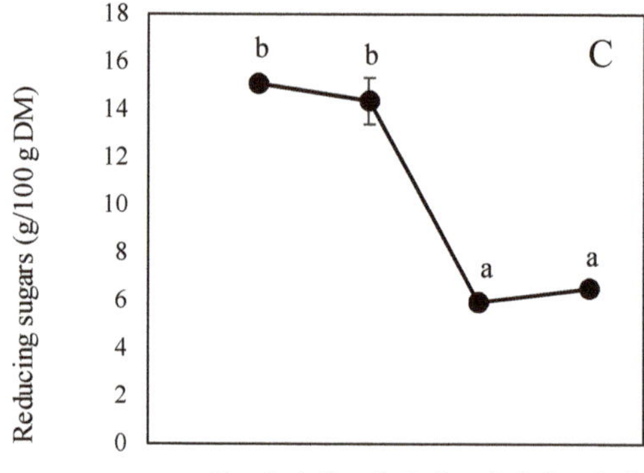

Figure 2. Development of acrylamide (**A**), asparagine (**B**) and reducing sugars (**C**) at different stages of panela production. All data expressed in dry matter (DM). Different letters indicate significant differences ($p < 0.05$).

3.2. HMF, Furfural and Browning

The HMF and furfural contents ranged from <LOQ to 2.37 mg/kg of FW and from <LOQ to 4.50 mg/kg of FW (Table 2) in sample 1 and sample 4 (panela), respectively (Figure 3A,B). When expressed according to dry matter, values ranged from <LOQ to 2.57 mg/kg for HMF and from <LOQ to 4.87 mg/kg for furfural. Both HMF and furfural are chemical markers of progress in the heat process [22]. However, the values are low considering the high sugar content (sucrose) seen in sugarcane and the high temperatures applied during the process. These values are similar to those reported in milk proteins (not detected (ND)–7.42 mg HMF/kg) [28,29] and infant formulas (not detected–14.2 mg HMF/kg and not detected–0.62 mg Furfural/kg), which do not have such a high sugar content or undergo such drastic thermal processes. HMF and furfural content in panela was quite low in comparison with other food matrices such as coffee (23.3–4112 mg HMF/kg) [30], dried fruit (25–2900 mg HMF/kg) and balsamic vinegar (316.4–35251.3 mg HMF/kg) [31]. Gómez-Narváez et al. [24] found average HMF and furfural values in panela blocks of 5.9 and 3.0 mg/kg, respectively, which is similar to those found in the present study. Since furfural is generated from pentoses and not from hexoses, it may be hypothesized that furfural is formed from the interconversion of HMF. This occurs as a result of strong heating conditions [32], which is supported by the fact that furfural was only found in the final product (block panela).

Measurement of absorbance at 420 nm in the soluble fraction of panela is a parameter that is used to monitor the extent of browning reactions [33]. Several compounds account for the absorbance seen at 420 nm. These include natural compounds present in the juice such as phytochemicals, and those formed during processing such as products of caramelization, the Maillard reaction or the oxidation of phenolic compounds. Browning ranged from 0.51 to 3.60 AU/g of FW (Table 2) for sample 1 and sample 4, respectively. Browning in panela is similar to that reported by Gómez-Narváez et al. [24] in block panela (mean 2.2 units/g). In a similar way to HMF and furfural, significant differences were observed ($p < 0.05$) when different stages of the process in DM were considered (Figure 3). In the case of browning, the greatest increase occurred in panela (1.6 times the value detected in sample 1) (Figure 3C). This was an expected result given the progress of the thermal treatment applied during different stages of block panela production. In accordance with the findings relating to other parameters analyzed in the present study, no outcomes were seen in relation to the progress of browning during the processing of raw sugar cane juice.

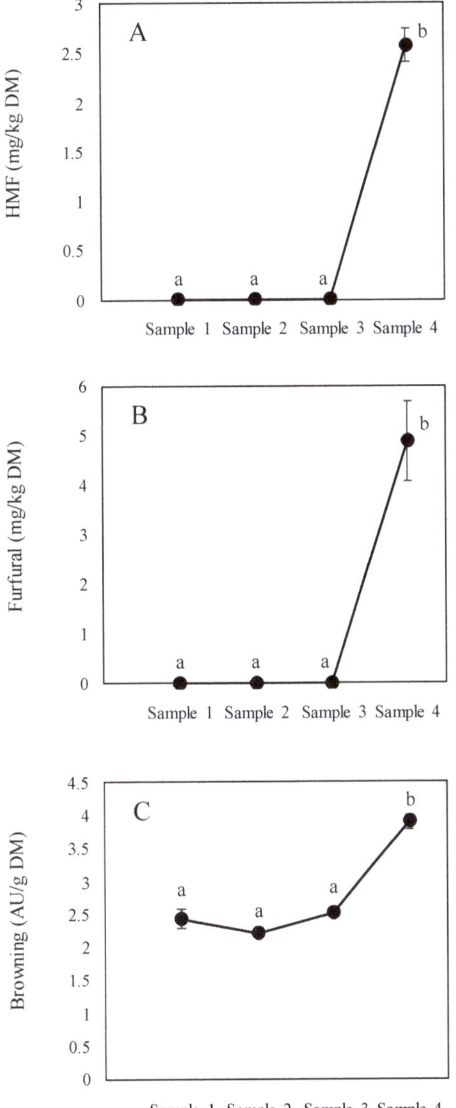

Figure 3. Development of HMF (**A**), furfural (**B**) and browning (420 nm) (**C**) during different stages of panela production. All data expressed in DM. Different letters indicate significant differences ($p < 0.05$).

3.3. Antioxidant Activity

TPC and ABTS ranged from 0.89 to 4.18 GAE/g of FW and from 5.08 to 29.70 μmol TE/g of FW, respectively (Table 2). When expressed according to dry matter, values ranged from 3.55 (sample 2) to 4.63 (sample 3) GAE/g and from 18.73 (sample 2) to 32.15 (sample 4) μmol TE/g for TPC and ABTS, respectively. A significant decrease in sample 2 vs. sample 1 was observed in both ABTS and TPC (Figure 4A,B), which may be due to the loss of natural antioxidants from cane juice. Subsequently, a significant increase was observed in sample 3 ($p < 0.05$) in relation to samples 1 and 2, whilst no significant differences ($p > 0.05$) were observed between sample 3 and the panela. This suggests that antioxidant compounds derived from the heat treatment are generated mainly during the concentration stage and remain constant in the final product (block panela). There are no reports in the literature regarding the evolution of antioxidant activity measured as the TPC and ABTS during panela production. The values found in the present study are within the range reported by Gómez-Narváez et al. [24] for block and granulated panela samples (1.1–6.2 mg of GAE/g and 12.7–50.5 µmol TE/g). However, lower values of the TPC (0.26 mg GAE/g) have been reported by Payet et al. [34], whereas much higher values (165–321 mg GAE/g) have been described by Lee et al. [27]. These differences may be due to different varieties of cane being used, alongside different panela processes, antioxidant extraction methods and measurement protocols. According to Payet et al. [34], antioxidant phenolics and flavonoids from the sugarcane stalk are retained in brown sugar during the non-centrifugation procedure. On the other hand, the high temperatures applied during the evaporation process promote non-enzymatic browning reactions, the formation of dark-colored substances with antioxidant activity and greater accessibility of phenolic compounds trapped in complex structures [35]. In this respect, significant correlations were observed between the ABTS results in relation to acrylamide ($r^2 = 0.8517$, $p = 0.0073$) and browning ($r^2 = 0.7230$, $p = 0.0427$). Figure 4 shows the evolution of antioxidant capacity measured as ABTS (4A) and the TPC (4B) during panela production (results expressed according to dry matter). A similar pattern is observed in both parameters, with a large increase being evident between sample 2 and sample 3, which is confirmed by a significant correlation between these samples ($r^2 = 0.9341$, $p = 0.0007$). This fact suggests the essential contribution of polyphenolic compounds to increase ABTS values.

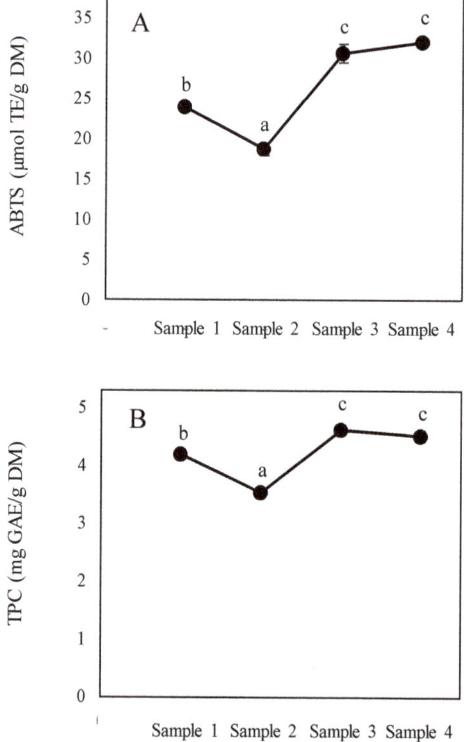

Figure 4. Evolution of antioxidant activity measured as ABTS (2,2′-azino-bis(3-ethylbenzothiazoline-6-sulfonic acid (**A**) and TPC (Total Phenolic Content) (**B**) during different stages of panela production. All data expressed in DM. Different letters indicate significant differences ($p < 0.05$).

4. Conclusions

The formation of acrylamide, HMF and furfural, the evolution of the antioxidant capacity and browning during different stages of panela production were evaluated by applying robust and validated analytical methods for the first time in the present study. Acrylamide occurs mainly at the stage of honey concentration and at the final stage of block panela manufacturing. Antioxidant compounds contributing to antioxidant capacity are mainly generated at the stage of honey concentration, whilst HMF, furfural and soluble compounds contributing to browning are primarily formed during the final stage of the process. The high temperatures applied during the evaporation process promote non-enzymatic browning reactions and the formation of dark-colored substances with antioxidant activity. Acrylamide formation during the final stage (block panela) takes place without interference from reducing sugars, possibly due to the presence of Maillard reaction intermediate compounds, which have been formed in previous stages of the process. These results provide greater insight into the identification of critical points in the process and suggest that mitigation strategies could be focused on the last stage of panela production, which would help panela producers and food safety bodies to control the formation of processing contaminants in this foodstuff.

Author Contributions: Conceptualization, F.J.M.; Methodology, M.M., C.D.-A., J.C.-C. and F.G.-N.; Software, M.M. and F.G.-N.; Validation, C.D.-A., J.C.-C. and F.J.M.; Formal Analysis, J.C.-C. and F.G.-N.; Investigation, M.M., C.D.-A. and F.J.M.; Resources, F.J.M.; Data Curation, M.M., C.D.-A., J.C.-C., F.G.-N. and F.J.M.; Writing—Original Draft Preparation, J.C.-C., F.G.-N.; Writing—Review and Editing, M.M., C.D.-A. and F.J.M.; Supervision, M.M., C.D.-A. and F.J.M.; Funding Acquisition, F.J.M. All authors have read and agreed to the published version of the manuscript.

Funding: This research was co-funded by the Spanish National Research Council (CSIC) (Spain) under project iCOOP (COOPB20288) and the Community of Madrid and European funding from FSE and FEDER programs (project S2018/BAA-4393, AVANSECAL-II-CM).

Acknowledgments: The authors would like to thank the University of Antioquia (UdeA).

Conflicts of Interest: The authors declare no conflict of interest.

References

1. Vargas Lasso, J.J.; Talero Pérez, Y.V.; Trujillo Suárez, F.A.; Caballero, L.R. Determinación de acrilamida en el procesamiento de la panela por cromatografía líquida Acrylamide Determintion in the Sugar Cane Juice Process by the Liquid Chromatography Technique. *Rev. Cienc. en Desarro.* **2014**, *5*, 99–105.

2. Osorio-Cadavid, G. *Manual Técnico: Buenas Prácticas Agrícolas -BPA- y Buenas Pácticas de Manufactura -BPM- en la Produccion de Caña y Panela*; FAO: Medellín, Colombia, 2007.

3. Food and Agriculture Organization Commodity Balances—Crops Primary Equivalent. Available online: http://www.fao.org/faostat/en/#data/BC/metadata (accessed on 21 March 2020).

4. De María, G. Panela: The natural nutritionalsweetener. *Agro Food Ind. Hi-Tech* **2013**, *24*, 44–48.

5. Jaffé, W. Health Effects of Non-Centrifugal Sugar (NCS): A Review. *Sugar Tech.* **2012**, *14*, 87–94. [CrossRef]

6. Durán-Castro, N. *Reingeniería Panelera*, 1st ed.; Buhos Editores: Tunja, Colombia, 2010; Available online: https://www.editorialprodumedios.com/agricola/269-reingenieria-panelera.html (accessed on 15 February 2020).

7. Kroh, L.W. Caramelisation in food and beverages. *Food Chem.* **1994**, *51*, 373–379. [CrossRef]

8. Park, J.; Kamendulis, L.; Friedman, M.; Klaunig, L. Acrylamide-Induced Cellular Transformation. *Toxicol. Sci.* **2002**, *65*, 177–183. [CrossRef]

9. Rice, J.M. The carcinogenicity of acrylamide. *Mutat. Res. Genet. Toxicol. Environ. Mutagen.* **2005**, *580*, 3–20. [CrossRef]

10. Friedman, M. Food browning and its prevention: An overview. *J. Agric. Food Chem.* **1996**, *44*, 631–653. [CrossRef]

11. Shapla, U.M.; Solayman, M.; Alam, N.; Khalil, M.I.; Gan, S.H. 5-Hydroxymethylfurfural (HMF) levels in honey and other food products: Effects on bees and human health. *Chem. Cent. J.* **2018**, *12*, 1–18. [CrossRef]

12. EFSA (European Food Safety Authority). Opinion of the scientific panel on food additives, flavourings, processing aids and materials in contact with food (AFC) on a request from the commission related to flavouring group evaluation 13: Furfuryl and furan derivatives with and without additional. *EFSA J.* **2005**, *215*, 1–73.

13. AOAC. *Official Method of Analysis of AOAC Intl*, 17th ed.; Association of Official Analytical Chemists: Gaithersburg, MD, USA, 2003.

14. Ministerio de Salud Panela Manual de Análisis. Available online: https://www.minsalud.gov.co/sites/rid/Lists/BibliotecaDigital/RIDE/IA/INS/ManualparaelanálisisdePanela.pdf (accessed on 22 March 2020).

15. Ayvaz, H. Rapid Assessment of Acrylamide and Its Precursors in Potato Tubers and Snacks by Infrared Spectroscopy. Ph.D. Thesis, The Ohio State University, Columbus, OH, USA, 2014.

16. Farkas, T.; Toulouee, J. Asparagine analysis in food products. *LC GC Eur.* **2003**, *21*, 14–16.

17. Mesias, M.; Delgado-Andrade, C.; Holgado, C.; Morales, F.J. Acrylamide content in French fries prepared in households: A pilot study in Spanish homes. *Food Chem.* **2018**, *260*, 44–52. [CrossRef] [PubMed]

18. Mesías, M.; Morales, F.J. Acrylamide in commercial potato crisps from Spanish market: Trends from 2004 to 2014 and assessment of the dietary exposure. *Food Chem. Toxicol.* **2015**, *81*, 104–110. [CrossRef] [PubMed]

19. Mesías, M.; Holgado, F.; Márquez-Ruíz, G.; Morales, F.J. Effect of sodium replacement in cookies on the formation of process contaminants and lipid oxidation. *LWT Food Sci. Technol.* **2015**, *62*, 633–639. [CrossRef]

20. Pérez-Jiménez, J.; Saura-Calixto, F. Literature Data May Underestimate the Actual Antioxidant Capacity of Cereals. *J. Agric. Food Chem.* **2005**, *53*, 5036–5040. [CrossRef] [PubMed]

21. Marfil, R.; Giménez, R.; Martínez, O.; Bouzas, P.R.; Rufián-Henares, J.A.; Mesías, M.; Cabrera-Vique, C. Determination of polyphenols, tocopherols, and antioxidant capacity in virgin argan oil (Argania spinosa, Skeels). *Eur. J. Lipid Sci. Technol.* **2011**, *113*, 886–893. [CrossRef]

22. Rufián-Henares, J.A.; Delgado-Andrade, C. Effect of digestive process on Maillard reaction indexes and antioxidant properties of breakfast cereals. *Food Res. Int.* **2009**, *42*, 394–400. [CrossRef]

23. Hoenicke, K.; Gatermann, R. Studies on the Stability of acrylamide in food during storage. *J. AOAC Int.* **2005**, *88*, 268–273. [CrossRef]

24. Gómez-Narváez, F.; Mesías, M.; Delgado-Andrade, C.; Contreras-Calderón, J.; Ubillús, F.; Cruz, G.; Morales, F.J. Occurrence of acrylamide and other heat-induced compounds in panela: Relationship with physicochemical and antioxidant parameters. *Food Chem.* **2019**, *301*, 125256. [CrossRef]

25. Knerr, T.; Lerche, H.; Pischetsrieder, M.; Severin, T. Formation of a novel colored product during the maillard reaction of d-glucose. *J. Agric. Food Chem.* **2001**, *49*, 1966–1970. [CrossRef]

26. Jaffé, W.R. Nutritional and functional components of non centrifugal cane sugar: A compilation of the data from the analytical literature. *J. Food Compos. Anal.* **2015**, *43*, 194–202. [CrossRef]

27. Lee, J.S.; Ramalingam, S.; Jo, I.G.; Kwon, Y.S.; Bahuguna, A.; Oh, Y.S.; Kim, M. Comparative study of the physicochemical, nutritional, and antioxidant properties of some commercial refined and non-centrifugal sugars. *Food Res. Int.* **2018**, *109*, 614–625. [CrossRef] [PubMed]

28. Contreras-Calderón, J.; Guerra-Hernández, E.; García-Villanova, B. Indicators of non-enzymatic browning in the evaluation of heat damage of ingredient proteins used in manufactured infant formulas. *Eur. Food Res. Technol.* **2008**, *227*, 117–124. [CrossRef]

29. Gómez-Narváez, F.; Medina-Pineda, Y.; Contreras-Calderón, J. Evaluation of the heat damage of whey and whey proteins using multivariate analysis. *Food Res. Int.* **2017**, *102*, 768–775. [CrossRef] [PubMed]

30. Contreras-Calderón, J.; Mejía-Díaz, D.; Martínez-Castaño, M.; Bedoya-Ramírez, D.; López-Rojas, N.; Gómez-Narváez, F.; Vega-Castro, O. Evaluation of antioxidant capacity in coffees marketed in Colombia: Relationship with the extent of non-enzymatic browning. *Food Chem.* **2016**, *209*, 162–170. [CrossRef]

31. Theobald, A.; Müller, A.; Anklam, E. Determination of 5-hydrox-ymethylfurfural in vinegar samples by HPLC. *J. Agric. Food Chem.* **1998**, *46*, 1850–1854. [CrossRef]

32. Wen, C.; Shi, X.; Wang, Z.; Gao, W.; Jiang, L.; Xiao, Q.; Deng, F. Effects of metal ions on formation of acrylamide and 5-hydroxymethylfurfural in asparagine-glucose model system. *Int. J. Food Sci. Technol.* **2016**, *51*, 279–285. [CrossRef]

33. Delgado-Andrade, C.; Morales, F.J.; Seiquer, I.; Navarro, M.P. Maillard reaction products profile and intake from Spanish typical dishes. *Food Res. Int.* **2016**, *43*, 1304–1311. [CrossRef]

34. Payet, B.; Sing, A.S.C.; Smadja, J. Assessment of antioxidant activity of cane brown sugars by ABTS and DPPH radical scavenging assays: Determination of their polyphenolic and volatile constituents. *J. Agric. Food Chem.* **2005**, *53*, 10074–10079. [CrossRef]
35. Asikin, Y.; Hirose, N.; Tamaki, H.; Ito, S.; Oku, H.; Wada, K. Effects of different drying-solidification processes on physical properties, volatile fraction, and antioxidant activity of non-centrifugal cane brown sugar. *LWT Food Sci. Technol.* **2016**, *66*, 340–347. [CrossRef]

 *foods*

Article

Effects of Thawing and Frying Methods on the Formation of Acrylamide and Polycyclic Aromatic Hydrocarbons in Chicken Meat

Jong-Sun Lee [1], Ji-Won Han [1], Munyhung Jung [2], Kwang-Won Lee [3] and Myung-Sub Chung [1,*]

[1] Department of Food Science and Technology, Chung-Ang University, 4726 Seodongdae-Ro, Daedeok-Myeon, Anseong-Si 17546, Korea; gpqls3873@naver.com (J.-S.L.); w_w3333@naver.com (J.-W.H.)
[2] Department of Food and Biotechnology, Graduate School, Woosuk University, Samnye-eup, Wanju-gun 55338, Korea; munjung@woosuk.ac.kr
[3] Department of Biotechnology, College of Life Sciences and Biotechnology, Korea University, Anam-Dong, Sungbuk-Gu, Seoul 02841, Korea; kwangwon@korea.ac.kr
* Correspondence: chungms@cau.ac.kr; Tel.: +82-31-670-3272

Received: 30 March 2020; Accepted: 23 April 2020; Published: 4 May 2020

Abstract: Air frying is commonly used as a substitute for deep-fat frying. However, few studies have examined the effect of air frying on the formation of potential carcinogens in foodstuffs. This study aimed to investigate the formation of acrylamide and four types of polycyclic aromatic hydrocarbons (PAHs) in air-fried and deep-fat-fried chicken breasts, thighs, and wings thawed using different methods, i.e., by using a microwave or a refrigerator, or by water immersion. The acrylamide and PAHs were analyzed by high-performance liquid chromatography–tandem mass spectrometry (HPLC-MS/MS) and gas chromatography–mass spectrometry (GC-MS), respectively. Deep-fat-fried chicken meat had higher acrylamide (n.d.–6.19 µg/kg) and total PAH (2.64–3.17 µg/kg) air-fried chicken meat (n.d.–3.49 µg/kg and 1.96–2.71 µg/kg). However, the thawing method did not significantly affect the formation of either acrylamide or PAHs. No significant differences in the acrylamide contents were observed among the chicken meat parts, however, the highest PAH contents were found in chicken wings. Thus, the results demonstrated that air frying could reduce the formation of acrylamide and PAHs in chicken meat in comparison with deep-fat frying.

Keywords: chicken; air frying; deep-fat frying; acrylamide; polycyclic aromatic hydrocarbons

---

1. Introduction

Although meat consumption has increased steadily in recent years, the consumption of low-fat, low-calorie, and high-protein chicken has grown significantly [1]. The number of chicken franchise stores in Korea has increased from 9000 in 2002 to 24,602 in 2018 [2]. Additionally, the domestic consumption of chicken has continued to grow from 7.5 kg per person per year in 2005 to 14.2 kg per person per year in 2018 [3]. The consumption of chicken in the USA has also increased from 39.2 kg per person per year in 2005 to 42.6 kg per person per year in 2018 [4]. In the European Union, the production of chicken has increased from 8169 billion tons per year in 2005 to 12,260 billion tons per year in 2018 [5]. Furthermore, the most frequently consumed meat cooked at high temperatures is fried chicken [6].

Generally, frying involves more rapid heat transfer in comparison with other cooking methods. The lowest temperature employed for frying is 140 °C, although fried food is typically cooked at temperatures between 175 °C and 195 °C [7]. High temperatures promote dehydration of the crust, oil intake, and the chemical reactions of various food constituents such as the denaturation of proteins and the caramelization of carbohydrates [8–10]. Moreover, the compounds produced via the

Maillard reaction during frying enhance the aroma, color, crust, and texture of the food but reduce its nutritional quality [11,12].

Many epidemiological studies have indicated that a high consumption of processed meats may increase the risk of cancer (e.g., breast, prostate, colorectal, ureter, and pancreatic) in humans because the high cooking temperatures employed in their production can result in high levels of carcinogenic compounds, such as acrylamide and polycyclic aromatic hydrocarbons (PAHs) [13–15]. Acrylamide, which is classified as a probable carcinogen to humans (Group 2A) by the International Agency for Research on Cancer (IARC) [16], is also produced during the frying of chicken [14]. Additionally, based on previous studies regarding the carcinogenicity, epidemiology, and mutagenicity of the PAH benzo[a]pyrene (B(a)P), it was categorized by the IARC as a Group 1 carcinogen, indicating that it is carcinogenic to humans [17]. Furthermore, PAHs are known to be endocrine disorder substances, which have long residual terms and exhibit high toxicities as carcinogens [18]. Formerly, the European Commission (EC) regulations required the use of B(a)P as a marker for the content of carcinogenic PAHs in foods [19]. However, as B(a)P was not always found in foods containing PAHs, in 2008, a group of four PAHs (PAH4) and a group of eight PAHs (PAH8) were proposed as better indicators based on data relating to occurrence and toxicity [20]. However, PAH8 measurements provided no additional advantage over PAH4 measurements. Based on comments from the European Food Safety Authority, in 2011, the EC expanded the scope of their regulations to include other food types and to add restrictions on PAH4 levels [21].

As the formation of carcinogenic compounds, such as acrylamide and PAHs, in chicken meat poses a significant risk to human health, studies are required to determine alternative cooking methods that produce healthier products without compromising the texture, flavor, taste attributes, and appearance [22–24]. For example, some previous studies have investigated limiting the formation of carcinogenic compounds through the use of precooking methods, such as microwave prethawing, predrying, and low-pressure frying [25–27]. An additional means to reduce the formation of carcinogenic compounds is the use of a different cooking method. For example, air frying is commonly used as a substitute for deep-fat frying. This method produces fried food using only a small quantity of fat through oil droplets spread in a hot air stream. Direct contact between the dispersion of oil droplets in hot air and the product inside a closed chamber provides constant heat transfer rates between the air and the food being fried. Thus, this technology permits a reduction of ~90% in the fat content of fried products [24]. Moreover, a 77% reduction in the acrylamide content of air-fried French fries has also been reported (30 µg/kg for air frying and 132 µg/kg for deep-fat frying [28]). Importantly, the air-fried food was crispy on the outside and moist on the inside, and the sensory properties of the final product were similar to those obtained after deep-fat frying. Furthermore, the smell of the food during cooking was less intense in comparison with other frying methods. For this reason, air fryer purchases have risen from 2% in 2014 to 38% in 2018 [29].

However, despite the number of studies conducted to evaluate the effect of hazardous compound formation in fried foods, few studies have examined the effect of air frying on the formation of acrylamide and PAHs in foodstuffs. Thus, we herein investigate the formation of acrylamide and four types of PAHs in chicken breasts, thighs, and wings fried by air frying and deep-fat frying after thawing using a microwave or a refrigerator, or by water immersion. Higher acrylamide and PAH contents were found in deep-fat-fried chicken meat than in air-fried chicken meat, but the thawing method did not significantly affect the formation of either acrylamide or PAHs. Thus, air frying is a promising method for reducing the formation of potentially hazardous compounds in chicken.

2. Materials and Methods

2.1. Raw Materials

Frozen skinless chicken breasts as well as thighs and wings with skin (Harim Co. Ltd., Iksan-si, South Korea) were purchased from a local food market (Jeonju, South Korea). Soybean oil (CJ

CheilJedang, Seoul, South Korea) was purchased from a local food market and used for frying the chicken samples.

2.2. Chemicals

Acrylamide (>99%) and $^{13}C_3$-acrylamide (>99%) were supplied by Cambridge Isotope Laboratories (Andover, MA, USA). Formic acid (>99%) was purchased from Sigma-Aldrich (St. Louis, MO, USA) and methanol was supplied by J.T. Baker (Phillipsburg, NJ, USA).

Strata-X (200 mg, 6 mL) and Bond Elut AccuCAT (200 mg, 3 mL) solid-phase extraction (SPE) cartridges were purchased from Phenomenex (Torrance, CA, USA) and Agilent Technologies (Santa Clara, CA, USA), respectively. Polyvinylidene fluoride (PVDF) syringe filters were purchased from Futecs Co. (Daejeon, South Korea).

Benzo(a)anthracene (B(a)A), benzo(b)fluoranthene (B(b)F), chrysene (CRY), and B(a)P standards as well as benzo(a)pyrene-d12 (B(a)P-d12) and chrysene-d12 (CRY-d12) internal standards (I.S.) were purchased from Supelco (Bellefonte, PA, USA). Organic solvents, including *n*-hexane, ethanol, and dichloromethane, were purchased from Burdick & Jackson (Muskegon, MI, USA). Potassium hydroxide (>85%) for alkali saponification was purchased from Showa Denko (Tokyo, Japan). Sodium sulfate anhydrous (>99%) for dehydration from Yakuri Pure Chemicals (Kyoto, Japan) and filter paper from Whatman (Maidstone, UK) were used for the dehydration process. Sep-Pak cartridges Agilent Technologies (Santa Clara, CA, USA) were used for SPE. Polytetrafluoroethylene (PTFE) membrane syringe filters from Whatman (Maidstone, UK) and 1 mL syringes from KOREAVACCINE (Ansan-Si, South Korea) were used for the filtration process.

All chemicals used for the determination of acrylamide and the PAHs were of analytical or high-performance liquid chromatography (HPLC) analytical grade.

2.3. Thawing Procedures

Chicken thigh, wing, and breast samples of approximately 100 g were employed for the thawing experiments. The frozen chicken samples were subjected to three different home-based thawing practices prior to frying. In accordance with the sanitation standard operating manual [30], the chicken was packed in a low-density polyethylene plastic bag (Cleanwrap, South Korea) for thawing by immersion in water and for refrigeration. Microwave thawing was conducted in a polypropylene microwave-specific container (LocknLock, South Korea). Specifically, the thawing practices used in this study were as follows: (i) Thawing in a microwave (KR-M201BWB; Winia Daewoo, South Korea) at 310 W for 3 min; (ii) thawing in a refrigerator (GC-114HCMP; LG Electronics, South Korea) at 4 °C for 24 h; and (iii) thawing by immersion in distilled water at 20 °C for 1 h (changing the water after 30 min). These thawing practices are suggested by the United States Department of Agriculture (USDA) as safe defrosting methods [31]. Frying was performed immediately after thawing to prevent contamination [30].

2.4. Frying Procedures

Deep-fat frying was performed in a domestic electric fryer (model DK-201; Delki, South Korea) with an adjustable temperature up to 190 °C, a 6 L capacity, and a nominal power of 2000 W. The temperature was controlled using a digital thermometer. Hot air frying was conducted using an HD9220 Air Fryer (Royal Philips Electronics N.V., Amsterdam, The Netherlands) with an adjustable temperature up to 200 °C, a 2.2 L capacity, and a nominal power of 1425 W. Approximately 100 g of each sample of the thawed chicken thighs, wings, and breasts were fried at the same time when the temperatures of both fryers reached 180 °C. Deep-fat frying was performed for 10 min using an oil volume of 3.6 L, and the oil was changed thrice for triplicate experiments. Air frying was performed for 35 min, and the samples were flipped once after 20 min without oil spray.

2.5. Sample Preparation

All fried chicken parts were cooled to room temperature (approximately 20 °C) after frying and the chicken bone was removed. The deboned skinless chicken breasts and the thighs and wings with skin were homogenized using a Ninja blender (Hai Xin Technology Co. Ltd., Shenzhen, China) for 2 min, and the homogenized samples were stored at 4 °C prior to analysis.

2.6. Determination of Acrylamide Content

The analytical method employed for acrylamide determination was based on The Ministry of Food and Drug Safety method with minor modifications [32]. Specifically, each homogenized chicken sample (1 g) was added to water (9 mL) and $^{13}C_3$ acrylamide (1 mL, 200 ng/mL) in a 50 mL tube. The tube was shaken at 250 rpm for 20 min, centrifuged (COMBI-514R; Hanil Scientific Inc., Gimpo-si, South Korea) at 3500 rpm for 5 min, and filtered through a 0.45 μm PVDF syringe filter. A Strata-X SPE column was conditioned with methanol (3.5 mL) followed by water (3.5 mL). An aliquot of the sample (1.5 mL) was introduced onto the cartridge followed by water (0.5 mL), and the eluent was discarded. Then, further water (1.5 mL) was passed through the column and the eluent was collected. A Bond Elut AccuCAT SPE column was conditioned using methanol (2.5 mL) followed by water (2.5 mL). Finally, after passing an aliquot (0.5 mL) of the eluate collected from the Strata-X SPE column through the preconditioned Bond Elut AccuCAT SPE column, an aliquot (1 mL) of the eluate from the Strata-X column was introduced onto the Bond Elut AccuCAT SPE column and was collected in 2 mL vials.

High-performance liquid chromatography–tandem mass spectrometry (HPLC-MS/MS) was performed using a Shimadzu 30A HPLC system coupled to a Shimadzu MS8040 MS/MS system (Shimadzu, Kyoto, Japan). The HPLC-MS/MS analytical conditions are outlined in Table 1. The multiple reaction monitoring (MRM) transitions for the quantitation of acrylamide and $^{13}C_3$ acrylamide (internal standard) were m/z 72 > 55 and m/z 75 > 58, respectively.

Table 1. Analytical conditions employed for high-performance liquid chromatography–tandem mass spectrometry (HPLC-MS/MS) analysis of acrylamide contents in chicken samples.

Item	Conditions
HPLC instrument	Shimadzu 30A
Column	Kinetex polar C18 (150 mm × 2.1 mm i.d., 2.6 μm particle size, Phenomenex)
Flow rate	0.3 mL/min
Oven temperature	26 °C
Injection volume	20 μL
Mobile phases	0.5% methanol in distilled water and 0.1% acetic acid
MS/MS instrument	Shimadzu MS8040
Ionization mode	Electrospray ionization, 5000 V, positive mode,
Detection mode	MRM mode
Desolvation gas, collision gas	N_2

MRM: multiple reaction monitoring.

A calibration curve for acrylamide was prepared in the range of 0.5–200 μg/L. The correlation coefficient was 0.999. Based on the calibration curve parameters, the limit of detection (LOD) and the limit of quantification (LOQ) were calculated as 0.186 μg/kg and 0.562 μg/kg, respectively.

2.7. Determination of PAH Content

The PAH content was determined using the method of Lee et al. [33] with modifications. Specifically, each homogenized chicken sample (1 g) was placed in a flat-bottomed 300 mL flask, and then spiked with an internal standard mixture (1 mL, 100 μg/kg of CRY-d12 and B(a)P-d12). The combined *n*-hexane phase was washed with distilled water (3 × 50 mL), dried over Na_2SO_4, and concentrated using a rotary evaporator (N-1110, EYELA, Seoul, South Korea) at 37 °C to give a final volume of 2 mL. The samples were then eluted using a Sep-Pak silica cartridge, which was activated with hexane (20 mL). The resulting solution was concentrated to dryness under N_2 gas at 40 °C, and the residue was redissolved in dichloromethane (1 mL). This solution was passed through a 0.45 μm PTFE membrane filter and transferred into a 2 mL amber screw cap vial. Finally, the PAH contents were analyzed using gas chromatography–mass spectrometry (GC-MS) following a slightly modified standard procedure. GC-MS analyses were performed using an Agilent Technologies 7890A gas chromatograph coupled to an Agilent Technologies 5977B mass spectrometer (Agilent Technologies, Santa Clara, CA, USA). The GC-MS conditions used to analyze the PAH contents are outlined in Table 2. The selected ions employed for the four PAHs and the standards in selected ion monitoring (SIM) mode were as follows: B(a)A (*m/z* **228**, 226, 229), B(b)F (*m/z* **252**, 250, 253), CRY (*m/z* **228**, 226, 229), B(a)P (*m/z* **252**, 250, 253), B(a)P-d12 (*m/z* **264**, 263, 265), and CRY-d12 (*m/z* **240**, 236, 241). The values in bold indicate the quantification ions.

Table 2. Gas chromatography–mass spectrometry (GC-MS) conditions for analysis of the four polycyclic aromatic hydrocarbons (PAHs) in chicken samples.

Item	Conditions
GC instrument	Agilent Technologies 7890A
Column	HP-5MS UI (30 m × 250 μm i.d., 0.25 μm film thickness)
Column oven temperature	80 °C (1 min) → 4 °C/min, 220 °C → 20 °C/min, 280 °C (10 min)
Post run	310 °C, 10 min
Flow rate	1.5 mL/min, helium
Injection mode	Splitless mode
Injection volume	1 μL
Injection temperature	320 °C
MS instrument	Agilent Technologies 5977B
Fragmentation mode	Electron impact at 70 eV
Detection mode	SIM mode

SIM: selected ion monitoring.

2.8. Statistical Analysis

The experimental data were evaluated using the analysis of variance (ANOVA). Using the SPSS software program (IBM Inc., Chicago, IL, USA), significant differences ($p < 0.05$) among the means were determined from triplicate analysis using Duncan's multiple range test. The differences between the means of the two frying methods were estimated using the *t*-test for independent samples. Values were considered significant at $p < 0.05$.

3. Results and Discussion

3.1. Acrylamide Contents of Air-Fried and Deep-Fat-Fried Chicken Meat Parts

Figure 1 depicts the retention times of the acrylamide standard and internal standard in the HPLC-MS/MS chromatograms of a typical chicken meat sample. The acrylamide contents of the air-fried and deep-fat-fried chicken breast, thigh, and wing samples after thawing in a microwave, in a refrigerator, and by immersion in water are presented in Table 3. The effects of the frying method on acrylamide formation were examined using the same thawing method and the same chicken meat part.

The acrylamide levels of the deep-fat-fried chicken meats ranged from n.d. (not detected) to 6.19 μg/kg, whereas those of the air-fried chicken meats ranged from n.d. to 3.49 μg/kg. Overall, the deep-fat-fried samples contained significantly higher acrylamide levels than the air-fried samples, with the exception of the thigh and wing parts thawed in a refrigerator. For this study, deep-fat frying was performed using soybean oil for 10 min, whereas air frying was conducted without an oil spray for 25 min. Despite the longer frying time, lower acrylamide levels were observed in the air-fried chicken meat samples. It has been proposed that acrylamide is formed from acrolein, which originates from lipid degradation (i.e., oxidized fatty acids or glycerol) [11]. Acrolein is produced when lipids are heated at high temperatures [34]. More specifically, acrolein forms acrylic acid through oxidation and can react to generate an intermediate acrylic radical. Both species can yield acrylamide in the presence of a nitrogen source and under favorable reaction conditions [35]. Similarly, the level of acrylamide formation increased during the heating of potatoes when oil was added [36].

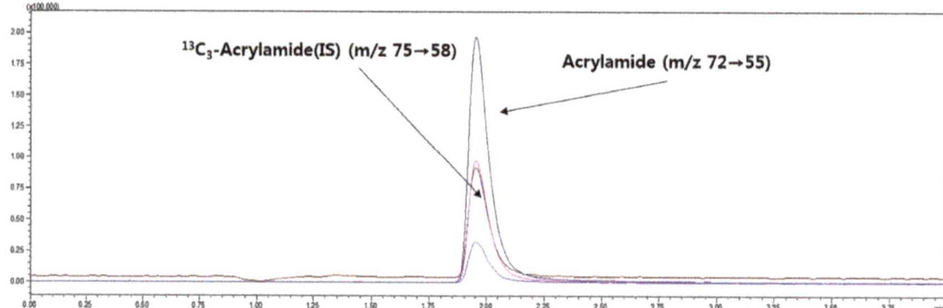

Figure 1. Typical high-performance liquid chromatography–tandem mass spectrometry (HPLC-MS/MS) chromatograms for the acrylamide standard (100 µg/kg) and the internal standard (20 µg/kg) in deep-fat-fried chicken wing samples after thawing by water immersion.

Table 3. Acrylamide levels of air-fried and deep-fat-fried chicken meats thawed using a microwave, a refrigerator, and water immersion.

Frying Method	Thawing Method	Chicken Part	Acrylamide Levels (µg/kg)[1]
Air frying	Microwave	Breasts	n.d.
		Thighs	n.d.
		Wings	3.49 ± 0.54^{BXa}
	Refrigerator	Breasts	n.d.
		Thighs	2.23 ± 1.50^{AXa}
		Wings	2.84 ± 0.68^{AXa}
	Water immersion	Breasts	n.d.
		Thighs	2.10 ± 0.55^{BXa}
		Wings	2.74 ± 0.20^{BXa}
Deep-fat frying	Microwave	Breasts	n.d.
		Thighs	2.85 ± 0.47^{AYb}
		Wings	4.91 ± 0.38^{AXa}
	Refrigerator	Breasts	2.52 ± 1.29^{AXb}
		Thighs	3.14 ± 0.95^{AXYab}
		Wings	4.91 ± 1.87^{AXa}
	Water immersion	Breasts	3.22 ± 0.82^{AXb}
		Thighs	4.62 ± 0.72^{AXb}
		Wings	6.19 ± 1.11^{AXa}

[1]: Each value represents the average of three independent repetitions ± standard deviation. A, B indicate statistically significant differences ($p < 0.05$) between the acrylamide levels of the same thawing methods and the same chicken meat parts where the frying method was varied. X, Y indicate statistically significant differences ($p < 0.05$) among the acrylamide levels of the same chicken meat parts and the same frying methods where the thawing method was varied. a, b indicate statistically significant differences ($p < 0.05$) among the acrylamide levels of the chicken meat parts where the same frying methods and same thawing methods were used. n.d. (not detected) indicates that the level was below the limit of detection (LOD).

3.2. Effect of Thawing Method on Acrylamide Formation

The effect of the thawing method (i.e., microwave, refrigerator, or water immersion) on the formation of acrylamide was examined using the same frying method and the same chicken meat part. No significant differences were observed ($p > 0.05$), except for the deep-fat-fried thigh meat thawed using a microwave. In this context, Erdoğdu et al. [37] reported that microwave precooking is an efficient way to decrease acrylamide formation in French fries by down-regulating the frying time and temperature. Additionally, Demirok and Kolsarıcı [14] used microwave pretreatment to reduce the levels of acrylamide from 95.15 to 94.13 µg/kg for coated chicken thighs and from 90.47 to 84.38 µg/kg for coated chicken wings.

3.3. Acrylamide Formation in Different Chicken Meat Parts

The formation of acrylamide in the chicken breasts, thighs, and wings thawed and fried using the same methods was examined. In general, independent of the frying method employed, the chicken wings contained the highest acrylamide contents, followed by the chicken thighs and the breast meat. No significant differences ($p > 0.05$) were observed among the air-fried chicken meat parts, whereas the deep-fat-fried chicken wings contained significantly higher acrylamide contents than the breast and thigh samples ($p < 0.05$). These results may be due to the different fat contents in the chicken meat parts. The EC established a recommended permitted value of 1000 µg/kg for residual acrylamide in potato crisps [38]; however, in our study, this value was not exceeded for any sample. Specifically, the acrylamide levels of the fried chicken samples produced herein were in the range of n.d.–3.22 µg/kg for the breast samples, n.d.–4.62 µg/kg for the thigh samples, and 2.74–6.19 µg/kg for the wing samples, which are extremely low levels. These results may be due to the use of fresh soybean oil in each frying experiment.

3.4. PAH Contents in Air-Fried and Deep-Fat-Fried Chicken Meat Samples

Figure 2 illustrates the GC-MS total ion chromatograms of a standard mixture of the four PAHs (PAH4, 10 µg/kg) and the internal standards (10 µg/kg). The PAH levels of the air-fried and deep-fat-fried chicken breast, thigh, and wing samples after thawing in a microwave, in a refrigerator, or by immersion in water are presented in Table 4. The results obtained for the two frying methods were compared for the same meat parts and thawing methods.

The sums of the four PAH levels for the deep-fat-fried chicken meat samples ranged from 2.60 to 3.17 µg/kg, whereas for the air-fried samples, these values ranged from 1.96 to 2.71 µg/kg. The deep-fat-fried chicken meats exhibited significantly higher ($p < 0.05$) total PAH levels than the air-fried chicken meats. Methyl linoleate is known to produce the highest levels of PAHs, followed by methyl oleate and methyl stearate, and it has been concluded that a greater extent of PAH formation occurs as the degree of unsaturation of the added lipids increases [39]. This behavior was attributed to unsaturated fatty acids being more prone to oxidation during heating [40]. It has previously been discussed that unsaturated fatty acids form cyclic monomers or dimers through polymerization [41]. According to Kostik et al. [42], the unsaturated fatty acid composition of soybean oil consists of oleic acid (28.5%, w/w), linoleic acid (49.5%), and linolenic acid (8%). Additionally, soybean oil should be susceptible to PAH formation because linoleic acid and linolenic acid are likely precursors of such lipid breakdown products [39]. Therefore, our results indicated that the unsaturated fatty acids present in the frying oil promoted PAH production.

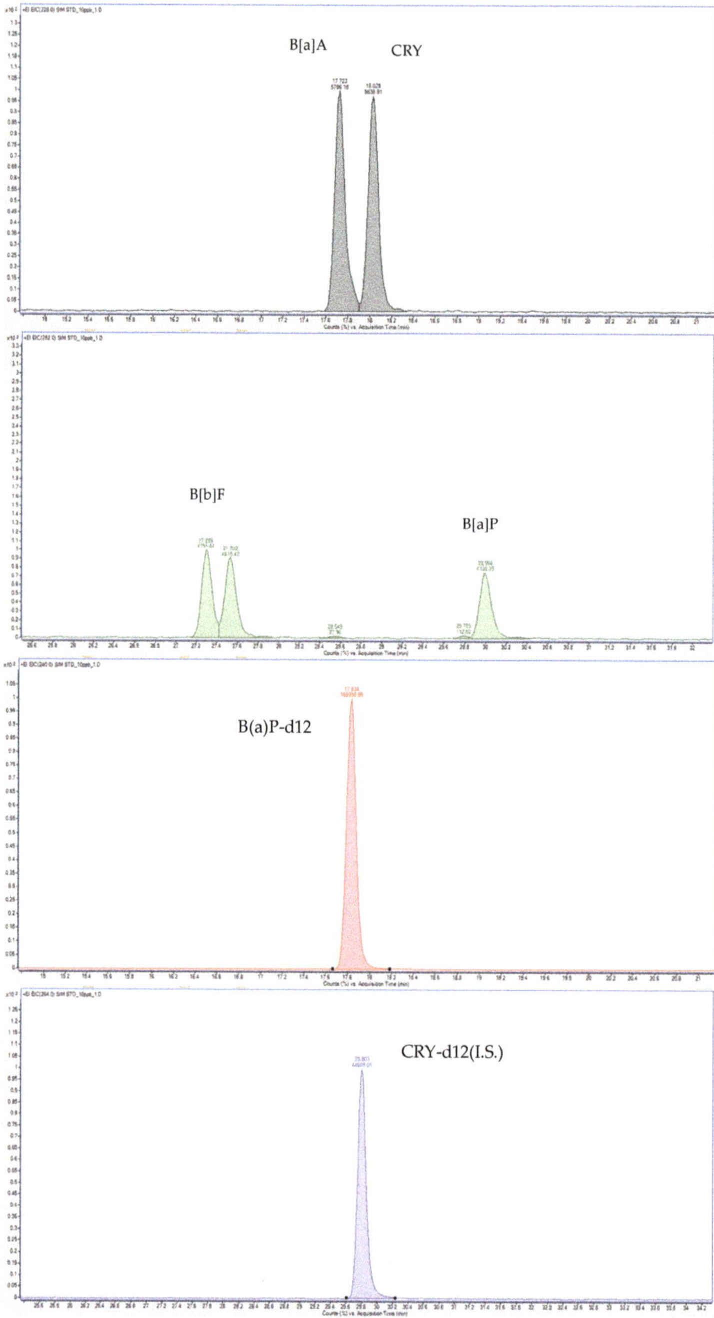

Figure 2. Gas chromatography–mass spectrometry (GC-MS) total ion chromatograms of a mixture of the four polycyclic aromatic hydrocarbons (PAHs) (10 µg/kg) and the B(a)P-d12 and CRY-d12 internal standards (10 µg/kg). Benzo(a)anthracene, B(a)A; benzo(b)fluoranthene, B(b)F; chrysene, CRY; benzo(a)pyrene, B(a)P; benzo(a)pyrene-d12, B(a)P-d12; and chrysene-d12, CRY-d12.

Table 4. Levels[1] (μg/kg) of the four PAHs in the air-fried and deep-fat-fried chicken breast, thigh, and wing samples thawed in a microwave, in a refrigerator, or by water immersion.

Frying Method	Thawing Method	Chicken Part	B(a)A	B(b)F	CRY	B(a)P	PAH4[2]
Air frying	Microwave	Breasts	0.27 ± 0.03^{BXb}	0.28 ± 0.05^{AXa}	0.94 ± 0.12^{BXb}	0.52 ± 0.03^{AXa}	2.00 ± 0.20^{BXb}
		Thighs	0.31 ± 0.02^{AXb}	0.24 ± 0.05^{AXa}	1.06 ± 0.01^{BYb}	0.54 ± 0.00^{BXa}	2.15 ± 0.07^{BXYb}
		Wings	0.43 ± 0.04^{AXa}	0.30 ± 0.05^{BXa}	1.43 ± 0.10^{AXa}	0.55 ± 0.05^{AXa}	2.71 ± 0.18^{BXa}
	Refrigerator	Breasts	0.26 ± 0.04^{AXa}	0.19 ± 0.02^{BYb}	0.95 ± 0.05^{AXb}	0.56 ± 0.08^{AXa}	1.96 ± 0.16^{BXb}
		Thighs	0.33 ± 0.03^{AXa}	0.20 ± 0.02^{BXb}	1.01 ± 0.05^{BYb}	0.55 ± 0.02^{BXa}	2.08 ± 0.07^{BYb}
		Wings	0.32 ± 0.04^{AYa}	0.35 ± 0.05^{AXa}	1.30 ± 0.08^{AXa}	0.54 ± 0.09^{AXa}	2.51 ± 0.18^{BXa}
	Water immersion	Breasts	0.29 ± 0.02^{BXab}	0.25 ± 0.04^{AXYa}	1.04 ± 0.08^{BXa}	0.51 ± 0.00^{BXa}	2.08 ± 0.10^{BXa}
		Thighs	0.24 ± 0.04^{BYb}	0.22 ± 0.03^{AXa}	1.22 ± 0.09^{AXa}	0.59 ± 0.05^{BXa}	2.26 ± 0.09^{BXa}
		Wings	0.32 ± 0.02^{BYa}	0.30 ± 0.06^{AXa}	1.10 ± 0.11^{BYa}	0.59 ± 0.09^{AXa}	2.31 ± 0.24^{BXa}
Deep-fat frying	Microwave	Breasts	0.42 ± 0.05^{AXa}	0.33 ± 0.07^{AXYab}	1.30 ± 0.09^{AXb}	0.59 ± 0.07^{AYa}	2.64 ± 0.14^{AXb}
		Thighs	0.39 ± 0.07^{AXa}	0.31 ± 0.06^{AXb}	1.44 ± 0.14^{AXab}	0.76 ± 0.04^{AXYa}	$2.90 \pm 0.2A^{Xab}$
		Wings	0.46 ± 0.06^{AXa}	0.42 ± 0.04^{AXa}	1.52 ± 0.06^{AXa}	0.76 ± 0.13^{AXa}	3.17 ± 0.15^{AXa}
	Refrigerator	Breasts	0.35 ± 0.06^{AXa}	0.40 ± 0.07^{AXa}	1.21 ± 0.24^{AXa}	0.64 ± 0.02^{AYa}	2.60 ± 0.21^{AXa}
		Thighs	0.40 ± 0.03^{AXa}	0.35 ± 0.03^{AXa}	1.40 ± 0.09^{AXa}	0.64 ± 0.02^{AYa}	2.79 ± 0.09^{AXa}
		Wings	0.40 ± 0.03^{AXa}	0.45 ± 0.06^{AXa}	1.40 ± 0.09^{AXa}	0.59 ± 0.01^{AXb}	2.84 ± 0.09^{AYa}
	Water immersion	Breasts	0.43 ± 0.05^{AXa}	0.27 ± 0.02^{AYb}	1.33 ± 0.05^{AXa}	0.79 ± 0.03^{AXa}	2.81 ± 0.13^{AXa}
		Thighs	0.40 ± 0.03^{AXa}	0.27 ± 0.04^{AXb}	1.33 ± 0.01^{AXa}	0.78 ± 0.10^{AXa}	2.78 ± 0.08^{AXa}
		Wings	0.46 ± 0.03^{AXa}	0.36 ± 0.03^{AXa}	1.38 ± 0.07^{AXa}	0.73 ± 0.09^{AXa}	2.93 ± 0.14^{AXYa}

[1]: Each value represents the average of three independent repetitions ± standard deviation. [2]: PAH4 is the sum of the benzo(a)anthracene (B(a)A), benzo(b)fluoranthene (B(b)F), chrysene (CRY), and benzo(a)pyrene (B(a)P) contents. [A,B] indicate statistically significant differences ($p < 0.05$) between the PAH levels of the same thawing methods and the same chicken meat parts where the frying method was varied. [X,Y] indicate statistically significant differences ($p < 0.05$) among the PAH levels of the same chicken meat parts and the same frying methods where the thawing method was varied. [a,b] indicate statistically significant differences ($p < 0.05$) among the PAH levels of the chicken meat parts where the same frying methods and same thawing methods were used.

3.5. Effect of Thawing Method on PAH Formation

The effect of the thawing method (i.e., microwave, refrigerator, or water immersion) on the formation of PAHs was investigated using the same frying method and same chicken meat parts. It was found that the thawing method had no significant effect ($p > 0.05$) on PAH formation in air-fried chicken meats, with the exception of the thigh meat thawed in a refrigerator. In the case of the deep-fat-fried samples, the total PAH levels in the chicken wings thawed in a refrigerator were significantly lower ($p < 0.05$) than those of the chicken wings thawed using either a microwave or water immersion.

Previous research confirmed that thawing in a refrigerator resulted in a lower drip loss of 0.62% among chicken breast samples [43]. Additionally, the presence of water is an important factor in PAH formation because it prevents incomplete combustion by providing oxygen during heating [44,45]. Therefore, it is likely that the lower quantities of PAHs formed in the chicken samples thawed using a refrigerator were due to reduced moisture loss.

3.6. PAH Formation in Different Chicken Meat Parts

The formation of PAHs was examined in the chicken breast, thigh, and wing samples that were thawed and fried using the same methods. Specifically, the air-fried chicken wings exhibited the highest PAH contents, followed by the chicken thigh and breast samples. The total PAH levels were in the range of 1.96–2.08 µg/kg for the chicken breasts, 2.08–2.26 µg/kg for the chicken thighs, and 2.31–2.71 µg/kg for the chicken wings. Similar trends were observed for the deep-fat-fried chicken samples. However, no significant differences ($p > 0.05$) were observed among the deep-fat-fried chicken meat samples, with the exception of the chicken breast thawed in the microwave. Lee et al. [33] reported average PAH contents of 0.60–0.76 µg/kg for chicken breasts, 0.94–1.14 µg/kg for chicken thighs, and 0.70–1.17 µg/kg for chicken wings. According to Koh and Yu [46], the lipid contents of chicken wings (14.9%) are higher than those of thighs (2.8%) and breasts (1.2%). Additionally, the most prevalent fatty acid is oleic acid (42.57%), followed by palmitic acid (27.5%) and linoleic acid (15.96%) in chicken meats. Additionally, it has been reported that the fat content of a sample is an important factor in determining the extent of PAH formation in grilled meat [47]. More specifically, PAH formation can occur upon the pyrolysis of organic matter, and the greatest concentrations of PAHs have been shown to arise from the pyrolysis of fat [48], which accounts for the higher levels of PAH formation observed in the chicken wing samples. The regulatory maximum level for B(a)P in smoked meat products is 2 µg/kg, whereas the maximum PAH4 level is 12 µg/kg [21]. In the present study, none of the chicken samples exceeded these maximum values; the B(a)P contents were 0.51–0.79 µg/kg for the chicken breast samples, 0.54–0.78 µg/kg for the chicken thigh samples, and 0.54–0.76 µg/kg for the chicken wing samples. Interestingly, these values were significantly lower than the level of 9.2 µg/kg in duck meat reported by Chen and Lin [49]. However, the B(a)P and PAH4 levels (max 0.79 and 3.17 µg/kg, respectively) were similar to previously reported levels for beef (0.59 and 4.32 µg/kg, respectively) [50]. Furthermore, the maximum B(a)P level fell within the range of n.d.–1.2 µg/kg previously reported for deep-fat-fried chicken breasts [51].

4. Conclusions

We herein reported our investigation into the effects of different frying methods (air frying and deep-fat frying) and thawing methods (microwave, refrigerator, and water immersion) of chicken meat parts (breasts, thighs, and wings) on the formation of acrylamide and PAHs. The air-fried samples exhibited lower acrylamide and total PAH contents than the deep-fat-fried samples due to the lower oil content used during the frying process. No significant differences were observed among the thawing methods and the chicken parts in terms of the acrylamide content. However, higher PAH contents were detected in the chicken wing samples in comparison with the chicken thigh and chicken breast samples, likely due to the higher fat content of the chicken wings. Additionally, the amounts of carcinogenic compounds detected in this study were lower than those reported in previous studies because no batter was employed, and the frying oil was relatively fresh. Overall, air frying reduced the formation of acrylamide and PAHs in comparison with deep-fat frying. Therefore, these results may be useful in determining the optimal frying method for chicken in terms of minimizing the formation of potentially hazardous substances.

Author Contributions: Conceptualization, J.-W.H.; methodology, M.-S.C., M.J., and K.-W.L.; investigation, J.-S.L. and J.-W.H.; writing—Original draft preparation, M.-S.C., J.-S.L., and J.-W.H.; writing—Review and editing, M.-S.C., M.J., and K.-W.L. All authors have read and agreed to the published version of the manuscript.

Funding: This research was supported by a grant (18162 MFDS 053) from Ministry of Food and Drug Safety in 2019.

Acknowledgments: This research was supported by the Chung-Ang University Graduate Research Scholarship in 2018.

Conflicts of Interest: The authors declare no conflict of interest.

References

1. Animal Husbandry and Nutrition. Available online: https://www.intechopen.com/books/animal-husbandry-and-nutrition/quality-of-chicken-meat (accessed on 15 April 2020).
2. Korean Fair Trade Commission. 2019. Available online: http://www.ftc.go.kr/www/selectReportUserView.do?key=10&rpttype=1&report_data_no=8091 (accessed on 23 November 2019).
3. Ministry of Agriculture Food and Rural Affairs. Available online: http://kass.mafra.go.kr/kass/phone/kass.htm (accessed on 25 November 2019).
4. The National Chicken Council. Per Capita Consumption of Poultry and Livestock, 1960 to Forecast 2020, in Pounds. Available online: https://www.nationalchickencouncil.org/about-the-industry/statistics/per-capita-consumption-of-poultry-and-livestock-1965-to-estimated-2012-in-pounds/ (accessed on 14 April 2020).
5. USDA/FAS PSD Online. Available online: https://fas.usda.gov/psdonline/psdQuery.aspx (accessed on 15 April 2020).
6. Lee, J.-K.; Yoon, K.-S. A study of adult's consumption of cooked food with high heat. *J. Korean Soc. Food Sci. Nutr.* **2011**, *40*, 290–307. [CrossRef]
7. Aladedunye, F.A.; Przybylski, R. Degradation and nutritional quality changes of oil during frying. *J. Am. Oil Chem. Soc.* **2009**, *86*, 149–156. [CrossRef]
8. Foegeding, E.A.; Davis, J.P. Food protein functionality: A comprehensive approach. *Food Hydrocoll.* **2011**, *25*, 1853–1864. [CrossRef]
9. Gertz, C. Fundamentals of the frying process. *Eur. J. Lipid Sci. Technol.* **2014**, *116*, 669–674. [CrossRef]
10. Sengar, G.; Sharma, H.K. Food caramels: A review. *J. Food Sci. Technol.* **2014**, *51*, 1686–1696. [CrossRef] [PubMed]
11. Ibrahim, R.M.; Nawar, I.; Yousef, M.I.; El-Sayed, M.I.; Hassanein, A. Protective Role of Natural Antioxidants Against the Formation and Harmful Effects of Acrylamide in Food. *Trends Appl. Sci. Res.* **2019**, *14*, 41–55. [CrossRef]
12. Zhang, Q.; Saleh, A.S.; Chen, J.; Shen, Q. Chemical alterations taken place during deep-fat frying based on certain reaction products: A review. *Chem. Phys. Lipids* **2012**, *165*, 662–681. [CrossRef]
13. Rose, M.; Holland, J.; Dowding, A.; Petch, S.R.; White, S.; Fernandes, A.; Mortimer, D. Investigation into the formation of PAHs in foods prepared in the home to determine the effects of frying, grilling, barbecuing, toasting and roasting. *Food Chem. Toxicol.* **2015**, *78*, 1–9. [CrossRef]
14. Demirok, E.; Kolsarıcı, N. Effect of green tea extract and microwave pre-cooking on the formation of acrylamide in fried chicken drumsticks and chicken wings. *Food Res. Int.* **2014**, *63*, 290–298. [CrossRef]
15. Carrabs, G.; Marrone, R.; Mercogliano, R.; Carosielli, L.; Vollano, L.; Anastasio, A. Polycyclic aromatic hydrocarbons residues in Gentile di maiale, a smoked meat product typical of some mountain areas in Latina province (Central Italy). *Ital. J. Food Saf.* **2014**, *3*, 102–104. [CrossRef]
16. International Agency for Research on Cancer (IARC). Some industrial chemicals. In *IARC Monographs on the Evaluation of Carcinogenesis Risks to Humans*; IARC: Lyon, France, 1994; Volume 60, pp. 435–453.

17. International Agency for Research on Cancer (IARC). Polynuclear aromatic compounds, part 1, chemical, environmental, and experimental data. In *IARC Monographs on the Evaluation of Carcinogenesis Risks to Humans*; IARC: Lyon, France, 1983; Volume 32, pp. 33–451.

18. Teruel, M.D.R.; Gordon, M.; Linares, M.B.; Garrido, M.D.; Ahromrit, A.; Niranjan, K. A comparative study of the characteristics of french fries produced by deep fat frying and air frying. *J. Food Sci.* **2015**, *80*, E349–E358. [CrossRef] [PubMed]

19. European Commission (EC). European Commission Regulation (EC) No 1881/2006 of 19 December 2006 Setting Maximum Levels for Certain Contaminants in Foodstuffs (Text with EEA relevance). *OJEU* **2006**, *364*, 5–24.

20. European Food Safety Authority. EFSA Scientific Opinion of the Panel on Contaminants in the food chain on a request from the European Commission on polycyclic aromatic hydrocarbons in food. *EFSA J.* **2008**, *724*, 1–114.

21. Commission Regulation (EU). Available online: https://op.europa.eu/en/publication-detail/-/publication/6a58ffa2-7404-4acf-b1df-298f611f813d/language-en (accessed on 15 April 2020).

22. Basuny, A.M.; Arafat, S.M.; Ahmed, A.A. Vacuum frying: An alternative to obtain high quality potato chips and fried oil. *Banats J. Biotechnol.* **2012**, *3*, 22–30.

23. Andrés-Bello, A.; García-Segovia, P.; Martínez-Monzó, J. Vacuum frying: An alternative to obtain high-quality dried products. *Food Eng. Rev.* **2011**, *3*, 63–78. [CrossRef]

24. Andrés, A.; Arguelles, Á.; Castelló, M.L.; Heredia, A. Mass transfer and volume changes in French fries during air frying. *Food Bioproc. Technol.* **2013**, *6*, 1917–1924. [CrossRef]

25. Tuta, S.; Palazoğlu, T.K.; Gökmen, V. Effect of microwave pre-thawing of frozen potato strips on acrylamide level and quality of French fries. *J. Food Eng.* **2010**, *97*, 261–266. [CrossRef]

26. Dueik, V.; Moreno, M.C.; Bouchon, P. Microstructural approach to understand oil absorption during vacuum and atmospheric frying. *J. Food Eng.* **2012**, *111*, 528–536. [CrossRef]

27. Cruz, G.; Cruz-Tirado, J.; Delgado, K.; Guzman, Y.; Castro, F.; Rojas, M.L.; Linares, G. Impact of pre-drying and frying time on physical properties and sensorial acceptability of fried potato chips. *Food Sci. Technol.* **2018**, *55*, 138–144. [CrossRef]

28. Sansano, M.; Juan-Borrás, M.; Escriche, I.; Andrés, A.; Heredia, A. Effect of pretreatments and air-frying, a novel technology, on acrylamide generation in fried potatoes. *J. Food Sci.* **2015**, *80*, T1120–T1128. [CrossRef]

29. Chosunbiz. Available online: https://biz.chosun.com/site/data/html_dir/2019/02/21/2019022100937.html (accessed on 25 November 2019).

30. The Ministry of Food and Drug Safety. Available online: http://www.prism.go.kr/homepage/origin/retrieveOriginDetail.do;jsessionid=259A3D50EBB62ADE8CC47A32B02495C6.node02?cond_research_name=&cond_research_start_date=&cond_research_end_date=&cond_organ_id=1471000&research_id=1470000-200700053&pageIndex=24&leftMenuLevel=120 (accessed on 20 November 2019).

31. United States Department of Agriculture. The Big Thaw—Safe Defrosting Methods for Consumers. Available online: https://www.fsis.usda.gov/wps/portal/fsis/topics/food-safety-education/get-answers/food-safety-fact-sheets/safe-food-handling/the-big-thaw-safe-defrosting-methods-for-consumers/ct_index/!ut/p/a1/jZFRT8IwFIV_zR5LO4dk-LYsMYCySYgy9kLKdtc22VrSXpz66y3zCQNK-9R7zpd77yktaUFLzd-V4KiM5u3pXU52bMUm4TRli3waPrJ59rbKn9KUxet7b9j-YciiG_krJ2H_8YsbGtzZZboUtDxwlETpxtBCABKuXQ_W0aIxpiaON4CfpOEVEicB0AunGhlUyXXdKi1ogRLIXgmCkvcDRGporHHoVdIBSlM7z1hSGe2O3dCgwp3SNXzQDS3P52Whv_MsWo9niyxi-fi34UKgP4brifllRGv2w-9tE72PYr-7hQYs2NHR-rJEPLiHgAWs7_uRMEa0MKpMF7BLiPTL0eLcSQ_da_H1nMyYeuk2sUu-AT1hIAs!/#1 (accessed on 25 November 2019).

32. The Ministry of Food and Drug Safety. Available online: http://www.law.go.kr/LSW/admRulInfoP.do?admRulSeq=2000000012210 (accessed on 20 November 2019).

33. Lee, Y.-N.; Lee, S.; Kim, J.-S.; Patra, J.K.; Shin, H.-S. Chemical analysis techniques and investigation of polycyclic aromatic hydrocarbons in fruit, vegetables and meats and their products. *Food Chem.* **2019**, *277*, 156–161. [CrossRef] [PubMed]

34. Umano, K.; Shibamoto, T. Analysis of acrolein from heated cooking oils and beef fat. *J. Agric. Food Chem.* **1987**, *35*, 909–912. [CrossRef]

35. Yasahura, A.; Tanaka, Y.; Hengel, M.; Shibamoto, T. Gas chromatographic investigation of acrylamide formation in browning model system. *J. Agric. Food Chem.* **2003**, *51*, 3999–4003. [CrossRef] [PubMed]

36. Tareke, E. Identification and Origin of Potential Background Carcinogens: Endogenous Isoprene and Oxiranes, Dietary Acrylamide. Ph.D. Thesis, Stockholm University, Stockholm, Sweden, 2003.
37. Belgin Erdoğdu, S.; Palazoğlu, T.K.; Gökmen, V.; Şenyuva, H.Z.; Ekiz, H.İ. Reduction of acrylamide formation in French fries by microwave pre-cooking of potato strips. *J. Sci. Food Agric.* **2007**, *87*, 133–137. [CrossRef]
38. Commission Recommendation (EC). Investigations into the levels of acrylamide in food. *OJEU* **2013**, *50*, L301/15.
39. Chen, B.H.; Chen, Y.C. Formation of polycyclic aromatic hydrocarbons in the smoke from heated model lipids and food lipids. *J. Agric. Food Chem.* **2001**, *49*, 5238–5243. [CrossRef]
40. Cossignani, L.; Giua, L.; Simonetti, M.S.; Blasi, F. Volatile compounds as indicators of conjugated and unconjugated linoleic acid thermal oxidation. *Eur. J. Lipid Sci. Technol.* **2014**, *116*, 407–412. [CrossRef]
41. Ezechi, E.H.; Affam, A.C.; Muda, K. *Handbook of Research on Resource Management for Pollution and Waste Treatment*, 1st ed.; IGI Global: Hershey, PA, USA, 2019; p. 29.
42. Kostik, V.; Memeti, S.; Bauer, B. Fatty acid composition of edible oils and fats. *J. Hyg. Eng. Des.* **2013**, *4*, 112–116.
43. Benli, H. Consumer attitudes toward storing and thawing chicken and effects of the common thawing practices on some quality characteristics of frozen chicken. *Asian-Australas. J. Anim. Sci.* **2016**, *29*, 100–108. [CrossRef]
44. Olatunji, O.S.; Fatoki, O.S.; Opeolu, B.O.; Ximba, B.J. Determination of polycyclic aromatic hydrocarbons [PAHs] in processed meat products using gas chromatography–Flame ionization detector. *Food Chem.* **2014**, *156*, 296–300. [CrossRef]
45. Min, S.; Patra, J.K.; Shin, H.-S. Factors influencing inhibition of eight polycyclic aromatic hydrocarbons in heated meat model system. *Food Chem.* **2018**, *239*, 993–1000. [CrossRef] [PubMed]
46. Koh, H.-Y.; Yu, I.-J. Nutritional analysis of chicken parts. *J. Korean Soc. Food Sci. Nutr.* **2015**, *44*, 1028–1034. [CrossRef]
47. Fretheim, K. Polycyclic aromatic hydrocarbons in grilled meat products—A review. *Food Chem.* **1983**, *10*, 129–139. [CrossRef]
48. Bartle, K.D. Analysis and occurrence of PAHs in food. In *Food Contaminants: Sources and Surveillance*, 1st ed.; Creaser, C., Purchase, R., Eds.; Royal Society of Chemistry: Cambridge, UK, 1991; pp. 41–60.
49. Chen, B.; Lin, Y. Formation of polycyclic aromatic hydrocarbons during processing of duck meat. *J. Agric. Food Chem.* **1997**, *45*, 1394–1403. [CrossRef]
50. Bogdanović, T.; Pleadin, J.; Petričević, S.; Listeš, E.; Sokolić, D.; Marković, K.; Ozogul, F.; Šimat, V. The occurrence of polycyclic aromatic hydrocarbons in fish and meat products of Croatia and dietary exposure. *J. Food Compost. Anal.* **2019**, *75*, 49–60. [CrossRef]
51. Aaslyng, M.D.; Duedahl-Olesen, L.; Jensen, K.; Meinert, L. Content of heterocyclic amines and polycyclic aromatic hydrocarbons in pork, beef and chicken barbecued at home by Danish consumers. *Meat. Sci.* **2013**, *93*, 85–91. [CrossRef]

Article

Industrial Strategies to Reduce Acrylamide Formation in Californian-Style Green Ripe Olives

Daniel Martín-Vertedor [1,2,]*, Antonio Fernández [1], Marta Mesías [3], Manuel Martínez [2,4], María Díaz [5] and Elisabet Martín-Tornero [4]

[1] Technological Institute of Food and Agriculture (CICYTEX-INTAEX), Junta of Extremadura, Avda. Adolfo Suárez s/n, 06007 Badajoz, Spain; tonymagic13@gmail.com
[2] Research Institute of Agricultural Resources (INURA), Campus Universitario, Avda. de la Investigación s/n, 06071 Badajoz, Spain; mmcano@unex.es
[3] Institute of Food Science, Technology and Nutrition (ICTAN-CSIC), Jose Antonio Novais 10, 28040 Madrid, Spain; mmesias@ictan.csic.es
[4] Environmental Engineering Agronomic and Forestry Department, University of Extremadura, Avda. Adolfo Suárez s/n, 06007 Badajoz, Spain; emartintornero@gmail.com
[5] Official College of Pharmacists, C/Ramón Albarrán 15, 06002 Badajoz, Spain; mardiamig@gmail.com
* Correspondence: daniel.martin@juntaex.es; Tel./Fax: +34-924-012-664

Received: 4 August 2020; Accepted: 28 August 2020; Published: 31 August 2020

Abstract: Acrylamide, a compound identified as a probable carcinogen, is generated during the sterilization phase employed during the processing of Californian-style green ripe olives. It is possible to reduce the content of this toxic compound by applying different strategies during the processing of green ripe olives. The influence of different processing conditions on acrylamide content was studied in three olives varieties ("Manzanilla de Sevilla", "Hojiblanca", and "Manzanilla Cacereña"). Olives harvested during the yellow–green stage presented higher acrylamide concentrations than green olives. A significant reduction in acrylamide content was observed when olives were washed with water at 25 °C for 45 min (25% reduction) and for 2 h (45% reduction) prior to lye treatment. Stone olives had 21–26% higher acrylamide levels than pitted olives and 42–50% higher levels than sliced olives in the three studied varieties. When calcium chloride ($CaCl_2$) was added to the brine and brine sodium chloride (NaCl) increased from 2% to 4%, olives presented higher concentrations of this contaminant. The addition of additives did not affect acrylamide levels when olives were canned without brine. Results from this study are very useful for the table olive industry to identify critical points in the production of Californian-style green ripe olives, thus, helping to control acrylamide formation in this foodstuff.

Keywords: acrylamide reduction; table olives; sterilization; additives

1. Introduction

Table olives are one of the main pickled products prepared throughout the world. Three main processing technologies are used worldwide to produce table olives: Spanish style, Californian-style and natural olives. The "Spanish-style" technique includes debittering the olives by soaking them in diluted lye solutions, washing them to remove excess lye and partial fermentation in brine. In the "Californian-style" method, olives are immersed in lye solutions, with or without air bubbling to cause darkening via oxidation, and then packaged and sterilized using retorts. The "natural olive" technique involves soaking olives in brine where they are subjected to spontaneous fermentation [1,2]. In general, the aim of all of these processing methods is to remove the natural bitterness of this fruit caused by the glucoside oleuropein, improve the sensory characteristics of olives, and ensure that they are safe for consumption [3].

The Californian-style, including Californian-style black ripe olives and Californian-style green ripe olives, is one of the most commonly used procedures, and olives treated according to this technique are the most commercialized. Both types of olives are obtained through very similar processes, however California-style green ripe olives are produced fresh without being stored in brine and are not submitted to air oxidation in tanks, thus avoiding the loss of their green color. The process starts with olives harvested at the green–yellow stage and, as mentioned before, being subjected to successive treatments with diluted NaOH (lye) to remove the natural bitterness. Then, olives are washed several times with water to remove most of the residual lye and lower the pH to 7–8. In the case of green olives, ferrous gluconate is not added for darkening. Finally, olives are canned in mild salt brine and heat sterilized at temperatures >110 °C [4–6].

It has been demonstrated that acrylamide is generated during the sterilization phase in California-style green ripe olives. Acrylamide is a toxic compound classified as a probable human carcinogen by the International Agency for Research on Cancer [7]. It is mainly produced as a result of the reaction between the asparagine amino acid and reducing sugars through the Maillard reaction, although different mechanisms appear to be involved in the formation of acrylamide in table olives [5,8]. In 2015, the European Food Safety Authority (EFSA) confirmed that the presence of acrylamide in foods is a public health concern [9]. EFSA considers Californian-style table olives as a potential source of acrylamide since these foods contain similar or even higher levels to those found in other food products such as French fries, cereals, or coffee. However, acrylamide concentrations in table olives can vary widely between different commercial canned black ripe olives [10], with values from 243 to 1349 ng·g^{-1} being recorded. Recently, olives in brine have been included among the foods that the European Commission recommends for monitoring due to the presence of acrylamide [11]. Mechanisms involved in the formation of this contaminant in this foodstuff should, therefore, be investigated.

Casado and Montaño [6] showed only trace or negligible amounts of acrylamide in olives prior to the sterilization process, however, concentrations increased up to 1578 ng·g^{-1} following sterilization. Thus, acrylamide has been associated with high-temperature treatments in Californian-style olives [5,6,10,12,13]. Furthermore, the application of different intensities of thermal sterilization treatments produces modifications in the sensory profile of Californian-style olives, mainly because it contributes a different cooked defect intensity [14]. In addition to temperature, other processing conditions such as storage time, washing of the fruit after oxidation, darkening methods, olive harvest, or the presence of additives can affect the acrylamide concentration [5,6,15]. Specifically, Charoenprasert and Mitchell [5] reported that acrylamide decreased when stored for longer than 30 days. Reduced lye treatment and washing with water also resulted in a higher acrylamide concentration [6], whilst olives processed when exposed to air had higher acrylamide levels than olives processed without air oxidation. This fact explains the higher acrylamide content of Californian-style black ripe olives when compared with Californian-style green ripe olives [16]. The cultivar also has a large influence on the variability of acrylamide levels in this food, with "Manzanilla de Sevilla" being the olive variety with the highest concentrations of this contaminant [6,13,16].

Additives in the brine solution (amino acids, water-soluble vitamins, or sodium sulfite, amongst others) have been described to influence acrylamide formation during the processing of ripe olives. Consequently, some have been tested as inhibitors of acrylamide formation. Casado et al. [17] found that the addition of cysteine (50 mM) caused a 50% reduction in acrylamide content. Lysine, arginine, and glycine also showed a significant decrease, with percentages dropping by a range of between 24% and 27%. López-López et al. [15] reported that proline and sarcosine are the most potent acrylamide inhibitors, whilst glycine, ornithine, taurine, and γ-aminobutyric acid are also effective. Both authors indicated that sodium bisulfite was the major inhibitor of acrylamide formation, with a low impact on sensory quality. However, this additive is currently not permitted by the European regulation for table olives. On the other hand, the absence of calcium chloride in the brine solution appeared to decrease the levels of acrylamide in olives following sterilization [5]. Based on these results, Charoenprasert and Mitchell [5] suggested that modifications to traditional processing methods such as keeping olives in the brine solution for longer periods, decreasing oxygen exposure time, or reducing sterilization temperatures can reduce acrylamide levels in table olives.

Another strategy that has been proved to control acrylamide formation is adding phenolic compounds to the can, such as olive leaf extract mixed with hydroxytyrosol. These constitute food additives and are introduced after the elaboration process and prior to the sterilization phase [10,13,18]. Casado et al. [17] reduced acrylamide in Californian-style olives by adding different natural vegetables with high antioxidant properties. Specifically, blanched garlic reduced acrylamide content by 23%. In contrast, the same authors found that adding to the brine two phenolic compounds, hydroxytyrosol and 3,4-dihydroxyphenyl glycol, which are naturally present in olives does not significantly affect acrylamide levels.

In conclusion, there are several steps during olive manufacturing that can affect acrylamide formation. As far as we know, most research on acrylamide in table olives has been focused on Californian-style black ripe olives and few studies have been conducted on green ripe olives. For this reason, the aim of the present work was to study the effect of processing conditions on acrylamide formation in Californian-style green ripe olives. Aspects considered included the harvest moment, washing prior to the sterilization process, the use of additives, and the form of presentation (stoned, pitted, sliced, or without brine). To our knowledge, this is the first study to focus on this kind of food. Although the acrylamide content of Californian-style green ripe olives is lower than that of black ripe olives, results from the present study could be very useful for controlling acrylamide production during the industrial processes producing green ripe olives.

2. Materials and Methods

2.1. Samples

Olives (*Olea europaea* L.) of the "Manzanilla de Sevilla", "Hojiblanca", and "Manzanilla Cacereña" varieties (as known as "Sevillana", "Hojiblanca", and "Cacereña", respectively) were obtained from a cultivar that collaborates with the "La Orden-CICYTEX" Research Center (Badajoz, Spain). Olive groves were located in the "Vegas Bajas del Guadiana" region. Olives were handpicked in perfect sanitary conditions during the 2018/19 crop season.

The climate of the region can be described as Mediterranean and sees an average annual rainfall of 404 mm. The olive orchard contained 25-year-old olive trees (6×6 m^2). The soil at the experimental orchard was a sandy loam (depth 2 m). About 3500 cm^3 water/ha was applied (between the 15 May and the 18 November) via linear drip irrigation. Weeds were controlled with post-emergence herbicides and by applying no-tillage conditions.

The three olive varieties were harvested at the green maturation stage (maturity index (MI = 0) according to skin color and flesh evaluations proposed by Uceda and Frías [19]. Olive sampling was carried out in the morning, taking samples randomly from different parts of the central area of the olive tree. After harvesting, olives were immediately introduced into tanks with NaOH (Dirna, Valencia, Spain) at 0.3% (154 kg table olives/225 L NaOH, approximately) and transported to the factory. Following this, olives were processed according to Californian-style green ripe olive techniques [5,16] in a factory located in the Northwest of Spain. Olives were then subjected to lye treatment, immersing the samples in a 2.5% (*w/v*) sodium hydroxide solution. This process took place at a room temperature of 25 °C and lasted until NaOH penetrated the pit of the olive. This solution was then removed, and the olives were placed into fresh water. The pH of the olives was neutralized using lactic acid and carbon dioxide gas until the pH value of the final table olives and brine solutions was 4.0. Following this, table olives (150 g) were packed in cans with different presentation formats (stoned olives, pitted olives, and sliced olives). A brine solution containing sodium chloride (NaCl, Valdequímica, Barcelona, Spain) (2% *w/v*) was added to the cans. Olive cans did not contain calcium chloride (CaCl$_2$). Finally, olive cans were sterilized in an autoclave at 121 ± 3 °C for 30 min. This treatment is equivalent to the cumulative lethality or "F0 value" of 15 min. Thus, we make sure that this is the treatment duration required to reduce microorganisms at a specified temperature to ensure the inactivation of thermo-resistant spoilage bacterial spores [16].

2.2. Experimental Design

Five types of industrial-scale experiments were carried out. The diagram of the experimental design is shown in Figure 1. All treatments were administered in quintiplicate. Acrylamide was determined both in olives and brine for experiments 1 and 2, and in olives, brine, and non-liquid olives for experiments 3, 4, and 5.

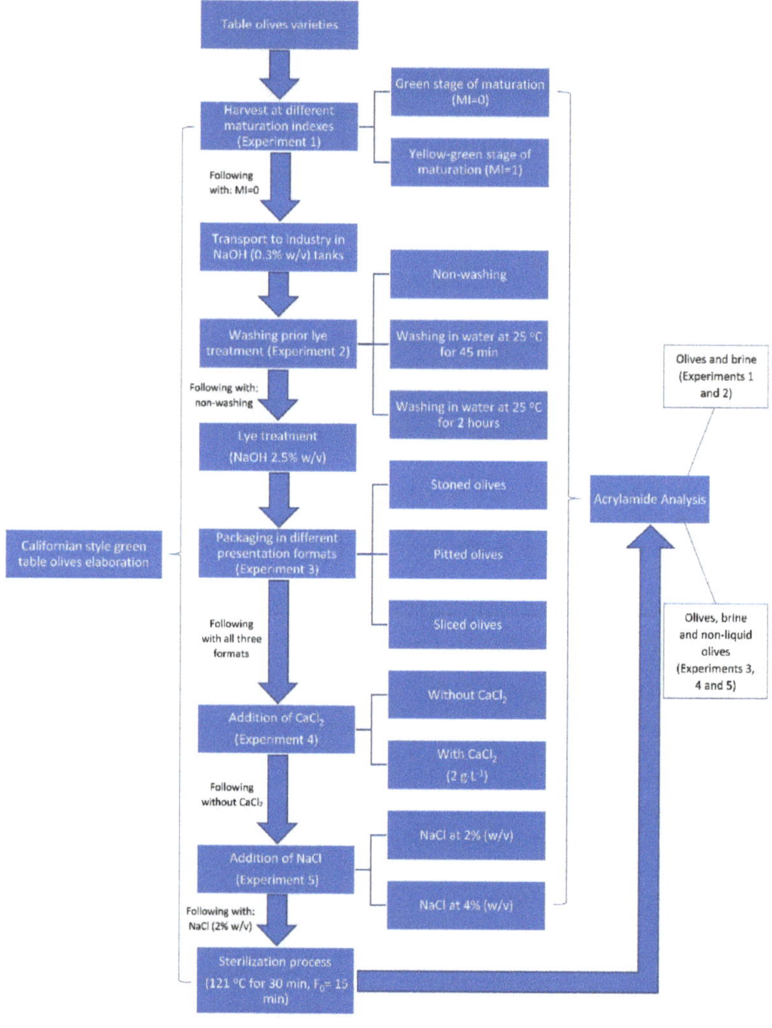

Figure 1. Diagram of the experimental design.

2.2.1. Experiment #1—Effect of the State of Ripeness of Olive Fruit

Two different maturation indexes (MI) of each olive variety were evaluated. Table olives were harvested during the same olive crop year at the green maturation stage (MI = 0) and at the yellow–green maturation stage (MI = 1). All samples were subjected to the same Californian-style treatment previously described.

2.2.2. Experiment #2—Effect of Washing Prior to Lye Treatment

Different washing treatments were applied to three varieties of table olives (MI = 0) prior to lye treatment: (i) non-washing, (ii) washing in water at 25 °C for 45 min (short washing), and (iii) washing in water at 25 °C for 2 h (long washing). In all cases, wash waters were removed, and table olives were sprayed for 5 min for superficial washing.

2.2.3. Experiment #3—Effect of the Presentation Format of Olives

Prior to packing, three presentation formats of the non-washed table olives obtained from experiment #2 were prepared: (i) stoned olives; (ii) pitted olives; and (iii) sliced olives. A model PSL-51 olive pitting and slicing machine (OFM Food Machinery, Seville, Spain) was used to obtain the different formats.

A batch of olives was put in brine (2% NaCl) while a second batch was canned without brine (non-liquid olives), with the latter being an innovation product. In this way, olives, brine, and non-liquid olives were evaluated in the three different formats (stone, pitted, and sliced olives).

2.2.4. Experiment #4—Effect of $CaCl_2$ Addition to the Brine Solution

After preparing the different presentation formats (stone, pitted, and sliced olives, with and without brine), the influence of adding $CaCl_2$ (Tetra Chemicals Europe, Helsingborg, Sweden) to the brine solution was studied. The following conditions were evaluated: (i) addition of $CaCl_2$ (2 g·L^{-1}) and (ii) no addition of $CaCl_2$.

2.2.5. Experiment #5—Effect of Different Concentrations of NaCl

Samples with the different presentation format (stone, pitted, and sliced olives) obtained from experiment #3 were introduced in different brine solutions: (i) with NaCl at 2% and (ii) with NaCl at 4%. $CaCl_2$ was not added to the samples.

2.3. Acrylamide Analysis in Olives and Brine Solutions

Samples of olives from the different experiments were crushed with a T-18 Basis Ultra-Turrax® Homogenizer device (IKA, Germany) in order to obtain a homogeneous olive paste. This was then stored at −80 °C whilst awaiting acrylamide analysis. Acrylamide was determined as described by Pérez-Nevado et al. [10]. Briefly, 2 g of table olives were stirred in 10 mL of distilled water, centrifuged at 1677 g at 4 °C for 10 min and filtered through a 0.22 μm nylon syringe filter (FILTER-LAB, Barcelona, Spain). Brine solution was only filtered. Filtered supernatants and brines were cleaned up using PCX (200 mg/3 mL) and PRP (60 mg/3 mL) column cartridges (Telos, Kinesis, Australia).

Sample extracts and calibration standards were analyzed on an Agilent Model 1200 Infinity LC high-performance liquid chromatograph (Agilent Technologies, Palo Alto, CA, USA) coupled with an Agilent Technologies triple quadrupole mass spectrometer, as outlined in the method proposed by Fernández et al. [20].

2.4. Statistical Analysis

SPSS 18.0 software was used for statistical analysis (SPSS Inc. Chicago, IL, USA). Data were expressed as mean values followed by the standard deviation (SD). One-way analysis of variance (ANOVA) followed by Tukey's multiple range test was used. The significance level was set at $p < 0.05$.

3. Results and Discussion

3.1. Effect of the State of Ripeness of Olive Fruit (Experiment #1)

In order to study the influence of different olive maturation indexes on acrylamide formation, levels of the contaminant were measured after submitting the three olive varieties at the green (MI = 0) and yellow–green (MI = 1) stages of maturation to Californian-style green olive processing. Results are shown in Figure 2.

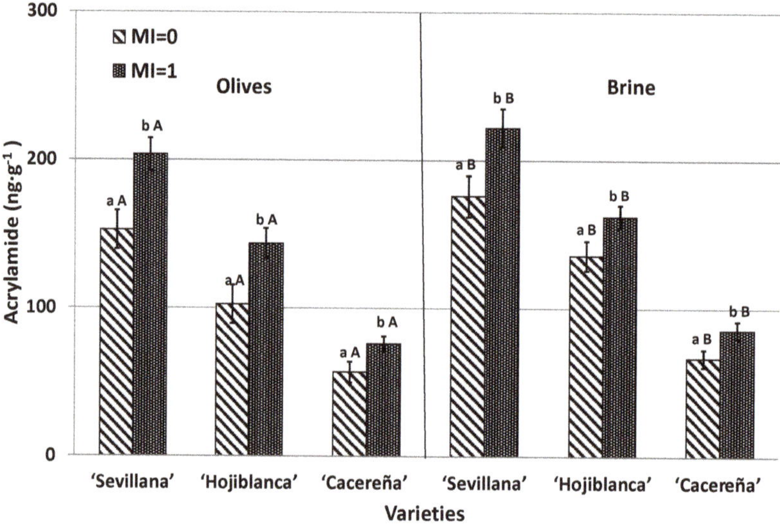

Figure 2. Acrylamide content (ng g^{-1}) in olives and brine obtained from different olive varieties harvested at two different olive maturation indexes (MI = 0 and MI = 1) and subjected to Californian-style green table olive processes. Results are expressed as the mean ± SD (standard desviation) of five sample replicates. Superscript indicates significant statistical differences between olives maturation stages for each olive variety (Tukey test, $p < 0.05$). Different capital letters in the same columns indicate significant statistical differences between olives and brine solution for each variety and maturation index (Tukey's test, $p < 0.05$).

Acrylamide was detected in all varieties studied. The "Sevillana" table olive variety exhibited the highest content of acrylamide, with mean values ranging from 153 ng·g^{-1} at the green stage of maturation (MI = 0) to 203 ng·g^{-1} at the yellow–green stage of maturation (MI = 1). Values ranging from 44 to 105 ng·g^{-1} have been reported by Charoenprasert and Mitchell [5] in "Sevillana" olives elaborated under the same Californian-style green ripe olive process. Lower concentrations were observed in the "Hojiblanca" variety, with mean values ranging from 103 to 144 ng·g^{-1} (33% and 29% lower than "Sevillana" in MI = 0 and MI = 1, respectively). The lowest levels were found in the "Cacereña" variety, with a mean concentration of 57 ng·g^{-1} in MI = 0 and of 76 ng·g^{-1} in MI = 1 (63% lower than "Sevillana" olives and around 45% lower than "Hojiblanca" olives for the two maturation indexes). Higher acrylamide concentrations in the "Sevillana" variety when compared with "Hojiblanca" and "Cacereña" olives have also been described in Californian-style black ripe olives [6,13]. Differences in acrylamide levels between the varieties may be due to the firmness of the fruit and/or the content of acrylamide precursors [6,10,13].

Yellow–green (MI = 1) table olives presented higher acrylamide concentrations than green olives (MI = 0) in the three olive varieties (32%, 39%, and 33% higher for "Sevillana", "Hojiblanca", and "Cacereña" olives, respectively) (Figure 2). This suggests that the most advanced stage of ripeness promotes the formation of this contaminant. This can be explained by the changes that occur in the composition and appearance of olives during the olive ripening process. Initially, the olive has an intense green color and becomes more yellowish during the maturation process. During this process, photosynthesis performed by the olive tree generates sugars and other compounds in the olive fruit, in this way increasing acrylamide precursors and, consequently, promoting greater acrylamide formation during sterilization treatment [10,16,21]. In addition, olives in the green–yellow stage contain a lower amount of phenolic compound than they do in the green stage of maturation, decreasing the protection given by these compounds against acrylamide formation [13].

Acrylamide levels in brine were significantly higher than that in olives for both indexes of maturation in the three studied varieties (Figure 2). This fact could be expected due to the diffusion of compounds from one matrix to another when olives are in brine. On the one hand, precursors of acrylamide in the fresh olives could diffuse into surrounding medium during brine storage [5] and, on the other hand, acrylamide generated after sterilization could also enter the brine. Acrylamide is a hydrosoluble and hydrophilic molecule and, therefore, when in aqueous mediums the chemical balance is expected to be displaced from the fruit to the brine. Similar results were found by Pérez-Nevado et al. [10] in Californian-style black ripe olives. These findings corroborate the information included in the recent publication of European Commission Recommendations [11]. This established that the high content of acrylamide in brined-stored food could be linked to the possible presence of acrylamide in brine.

According to our knowledge, this is the first study relating acrylamide formation and the ripeness state of olives. Considering the results, it is recommended that companies producing Californian-style olives acquire the raw fruits as early as possible when still in the earlier maturation state. In other words, the color of the olives should be as close as possible to green. In addition, due to the acrylamide presence in brine, consumers should remove brine before eating olives. It may also be advisable to add water to the olives for a few minutes to wash them and eliminate acrylamide from the brine. In this way, acrylamide exposure from table olives could be reduced.

3.2. Effect of Washing Prior to Lye Treatment (Experiment #2)

The acrylamide content of olives subjected to different washing processes prior to lye treatment is shown in Table 1. Significant differences were observed between the three conditions (non-washing, washing in water at 25 °C for 45 min (short washing), and washing in water at 25 °C for 2 h (long washing)). The lowest levels of acrylamide were observed in olives washed for a long time, followed by short-washed olives and, finally, non-washed olives. When compared with non-washed olives, 2 h washes reduced the formation of the contaminant by around 45% in the three varieties, with 45 min washes causing a reduction of around 25%. According to these results, it could be suggested that the cleaning step prior to lye treatment in lipid-rich foods removes acrylamide precursors from the olive fruit, subsequently decreasing contaminant formation. A similar mitigation strategy has been proposed for potatoes, in that it is recommended to wash or even soak fresh tubers before frying in order to reduce acrylamide precursors in the preparation of French fries or chips [11].

When olives were washed, brine also presented lower acrylamide concentrations (25%, 32%, and 22% lower in "Sevillana", "Hojiblanca", and "Cacereña" olives, respectively, when olives were washed during 45 min and 41%, 51%, and 36% lower for the same varieties when olives were washed for 2 h, when compared with levels in non-washed olives). This could be due to a lower diffusion of precursors from the fruits to the brine and lower diffusion of acrylamide from the sterilized olives to the brine.

Casado and Montaño [6] showed that olive fruits with a lower acrylamide concentration were those that underwent a washing process twice as long as that of comparison olives. These researchers indicated that long washing (24 h) following lye treatment reduced acrylamide content by approximately 50% in "Sevillana" olives (from 1340 to 680 ng·g^{-1};) and around 80% in the "Hojiblanca" variety (from 1130 to 200 ng·g^{-1}). These percentages are much higher than those obtained in the present study, probably due to the shorter washing time employed in our experiment but also due to the step in the process during which washing was carried out. These authors washed olives during the NaOH treatment, which favors greater diffusion and elimination of precursors. It is known that treatments with NaOH modify olive texture, making them less hard and more porous [10,22]. This promotes greater diffusion of compounds from the olives to the brine.

Table 1. Acrylamide content (ng·g^{-1}) of olives and brine after submitting olives to washing (non-washing, washing in water at 25 °C for 45 min (short washing), and washing in water at 25 °C for 2 h (long washing) after the sterilization process. Results are expressed as mean ± SD of five sample replicates. Superscript in the same row indicates statistically significant differences between washing processes (Tukey test, $p < 0.05$). Different capital letters in the same columns indicate statistically significant differences between olives and brine in each treatment (Tukey test, $p < 0.05$).

		Acrylamide (ng·g^{-1})		
	Samples	Non-Washing	Short Washing	Long Washing
"Sevillana"	Olives	210 ± 8 [c A]	161 ± 5 [b A]	124 ± 7 [a A]
"Sevillana"	Brine	233 ± 9 [c B]	175 ± 8 [b B]	137 ± 4 [a B]
"Hojiblanca"	Olives	151 ± 10 [c A]	108 ± 7 [b A]	84 ± 7 [a A]
"Hojiblanca"	Brine	184 ± 8 [c B]	125 ± 7 [b B]	90 ± 7 [a B]
"Cacereña"	Olives	79 ± 6 [c A]	58 ± 8 [b A]	44 ± 8 [a A]
"Cacereña"	Brine	84 ± 7 [c B]	65 ± 7 [b B]	54 ± 4 [a B]

In conclusion, industries should include a washing step prior to the elaboration process in order to reduce acrylamide precursors and, consequently, to decrease contaminant formation. According to the results of the present experiment, washing in water at 25 °C for 2 h could be an adequate mitigation strategy.

3.3. Effect of the Presentation Format of Olives on Acrylamide Content (Experiment #3)

The influence of the different table olive presentation formats (stone, pitted, and sliced olives) prior to packing, with and without brine, on acrylamide content in the three varieties is shown in Figure 3. For olives canned in brine, stone olives presented the highest concentrations in the three studied olive varieties, with levels 21–26% greater than pitted olives and 42–50% higher than sliced olives. Stone olives have a compact format enabling a smaller contact surface between the fruit and the liquid. This reduces the diffusion of both acrylamide and acrylamide precursors between the matrices, thus leading to the highest levels. In contrast, sliced olives have a greater contact surface with the brine solution throughout the different steps of the industrial elaboration process. This promotes a greater diffusion of acrylamide and acrylamide precursors from the olives to the brine. This fact explains why sliced presentations showed the lowest concentrations of this toxic substance. Our results agree with those of Casado and Montaño [6], verifying that oxidized sliced black olives pertained to the format with the lowest final acrylamide concentration following the sterilization process. As expected, pitted olives exhibited moderate values for acrylamide levels (Figure 3).

Similarly to previously conducted experiments, brine presented significantly higher concentrations than olives canned in brine and the trend according to the different format presentations was similar to those observed in the fruits: stone > pitted > sliced.

This trend was also noticed in olives canned without brine (non-liquid olives). Again, stone olives displayed the highest acrylamide concentration (20% higher than pitted olives and 45% higher than sliced olives). When compared with olives canned with brine, levels of the contaminant were significantly higher in samples canned without brine (non-liquid olives), with recorded values being as much as three times larger. No previous information has been found regarding the formation of acrylamide in olives canned without brine during the sterilization process. However, it could be deduced that the absence of a liquid medium in the can implies that there is no diffusion, of either the precursors or of acrylamide, to the aqueous medium and, therefore, all the acrylamide generated remains in the olives. In addition, when olives are canned with brine, the heat applied during sterilization is transmitted by convection and the thermal process occurs in a less aggressive way [10]. The absence of liquid could cause overheating in olives, leading to increased formation of acrylamide. For that reason, the consumption of olives with this presentation format, with more than 250 ng g^{-1} of acrylamide, should be controlled, especially among children and elderly consumers, in order to reduce a high acrylamide exposure and, consequently, to prevent possible health risks [5,18].

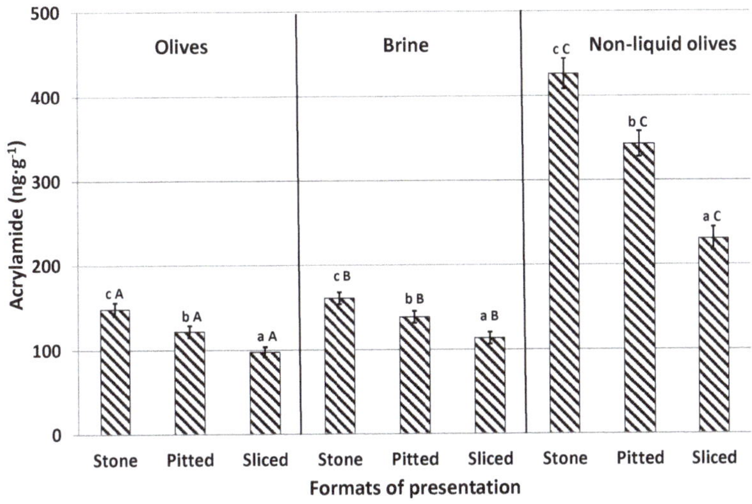

(**a**) Olives of the "Sevillana" variety elaborated such as Californian-style green ripe olive with different presentation formats.

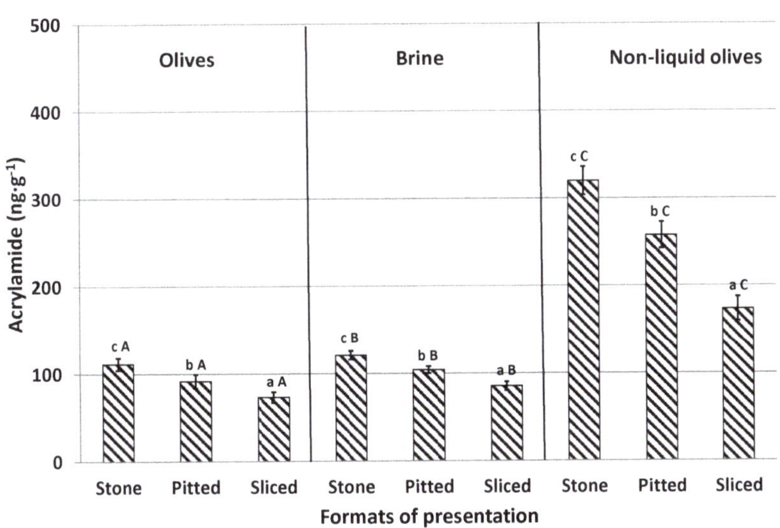

(**b**) Olives of the "Hojiblanca" variety elaborated such as Californian-style green ripe olive with different presentation formats.

Figure 3. *Cont.*

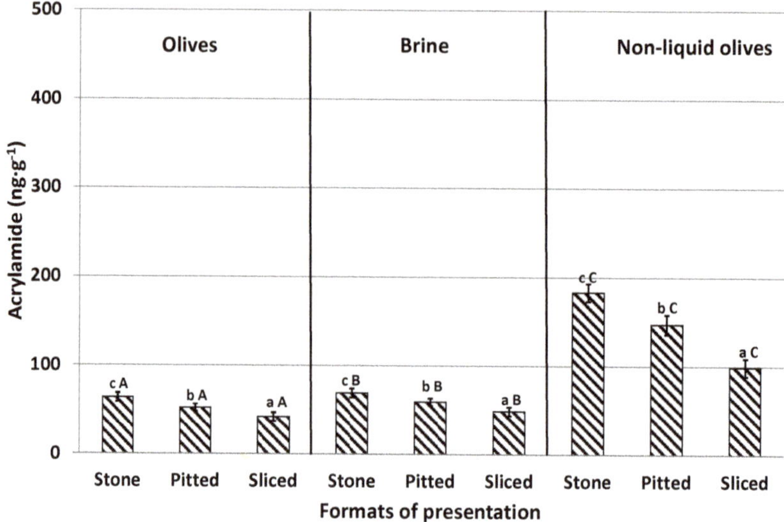

(**c**) Olives of the "Cacereña" variety elaborated such as Californian-style green ripe olive with different presentation formats.

Figure 3. Acrylamide content (ng g^{-1}) of olives (maturity index (MI = 0)) and brine with different presentation formats (stone, pitted, and sliced olives), canned with and without brine (non-liquid olives) after the sterilization process. Results are expressed as means ± SD of the five sample replicates. Superscript indicates significant statistical differences between presentation formats in each sample (Tukey test, $p < 0.05$). Different capital letters in the same columns indicate significant statistical differences between raw materials (olives, brine, and non-liquid olives) in each individual presentation format (Tukey´s test, $p < 0.05$).

3.4. Effect of CaCl$_2$ Addition to the Brine Solution (Experiment #4)

CaCl$_2$ is a firming agent, which is frequently added to the covering brine of packed table olives to improve olive firmness [23]. The influence of the addition or absence of CaCl$_2$ in the brine solution on acrylamide formation in the different presentation formats of table olives and after being submitted to the sterilization process is shown in Figure 4.

Olives canned with added CaCl$_2$ showed acrylamide levels approximately 20% higher than samples without CaCl$_2$. These higher levels coincide with the lower acrylamide concentrations of the brine in samples treated with CaCl$_2$. These results agree with those reported by Casado et al. [17] in an olive model system with olive juice obtained from "Hojiblanca" olives. They also coincide with those described by Charoenprasert and Mitchell [5] in Californian-style black ripe olives. These authors indicated that, in general, calcium ions help to maintain structural firmness and stability of the cell wall and the cellular turgor of the fruits that form cross links between pectin molecules. This strengthens plant cells and prevents their collapse. It is deduced that calcium reduces the transcription of acrylamide and acrylamide precursors between olives and brine despite these compounds being hydrophilic. The retention of acrylamide precursors inside the fruit promotes a higher formation of the contaminant, whilst the retention of generated acrylamide in sterilized olives increases its concentration in the final product.

The effect of adding calcium to samples canned without brine was not significant and similar concentrations were observed between olives treated and not treated with this additive. This was the case regardless of presentation format and olive variety. In this case, both acrylamide and acrylamide precursors could not migrate from olives to brine in cases where brine was not present. In summary, the presence of $CaCl_2$ does not affect acrylamide formation but can reduce the diffusion of acrylamide and acrylamide precursors from olives to brine when olives are canned in an aqueous environment. The addition of this additive to brine should be avoided in order to reduce the acrylamide content of olives after the sterilization process. However, industries must take into account that the firmness of the olives could be modified, and consumer's acceptance may be affected. For olives without brine, the use of $CaCl_2$ can be continued.

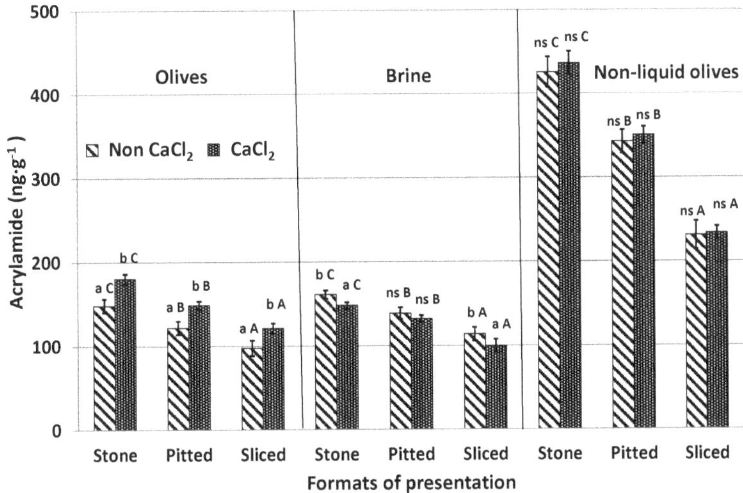

(**a**) Olives of the "Sevillana" variety elaborated such as Californian-style green ripe olive with different presentation formats and with or without the addition of CaCl₂ to the brine.

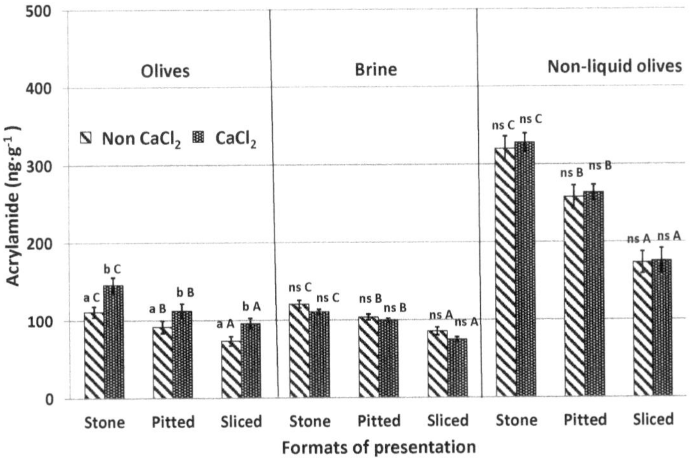

(**b**) Olives of the "Hojiblanca" variety elaborated such as Californian-style green ripe olive with different presentation formats and with or without the addition of CaCl₂ to the brine.

Figure 4. *Cont.*

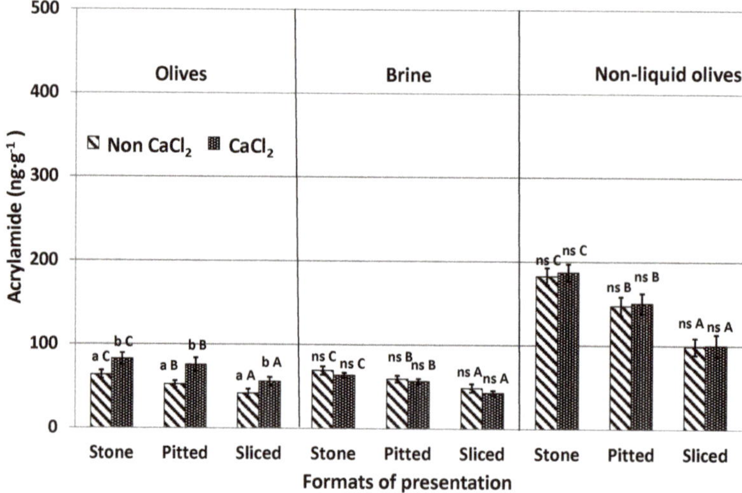

(**c**) Olives of the "Cacereña" variety elaborated such as Californian-style green ripe olive with different presentation formats and with or without the addition of CaCl₂ to the brine.

Figure 4. Acrylamide content (ng g^{-1}) of olives and brine obtained from different olive presentation formats (stones, pitted, sliced), with or without the addition of CaCl$_2$ to brine after the sterilization process. Results are expressed as means ± SD of five sample replicates. Superscript indicates statistically significant differences relating to CaCl$_2$ addition (Tukey test, $p < 0.05$). Different capital letters in the same columns indicate statistically significant differences between different olive presentation formats (Tukey test, $p < 0.05$). ns: no significant differences.

3.5. Effect of Different NaCl Concentrations (Experiment #5)

The influence of different NaCl concentrations in brine on acrylamide content in the different presentation formats of olives and varieties is shown in Table 2. NaCl is added during the processing of the table olive in order to give a salty taste to the final product. In this experiment, two different NaCl concentrations (2% and 4%) were added to the different olive samples and acrylamide content was analyzed in both the olives and the brine. Acrylamide concentration in olives in the three presentation formats was higher when 4% of NaCl was added to the brine. In contrast, acrylamide decreased in brine with a higher concentration of NaCl. Results are similar to those found in experiment #4 with the application of CaCl$_2$ but to a lesser extent. Although the objective of adding NaCl is different, resulting behavior suggests that this additive also indirectly strengthens the olive cellular membranes and, therefore, prevents the diffusion of acrylamide precursors from the olives to the brine. This gives rise to an increase in acrylamide content. In agreement with this finding, Fadda et al. [24] reported that olives canned in brine with higher NaCl percentages showed higher breaking force and hardness values than olives processed with lower NaCl contents.

Regarding the presentation format, when the percentage of NaCl increases, acrylamide formation increases to a lesser extent in stone olives, with greater increases seen in pitted and sliced samples. The greater contact surface in pitted and sliced olives probably allows NaCl to make the cell structures of olives harder, reducing the diffusion of acrylamide precursors from olives to brine and, consequently, increasing acrylamide formation in these formats when compared with stone olives.

Finally, the addition of salt led to no significant differences when olives were not canned with brine. This is an expected result because, as mentioned before, the absence of a liquid medium in the can prevents the diffusion of precursors and acrylamide to the aqueous medium and, as a result, all generated acrylamide remains in the olives. No bibliographic references have been found to confirm or disprove the results obtained in this experiment.

Table 2. Acrylamide content (ng·g^{-1}) of olives and brine after NaCl addition to brine at different concentrations (2% and 4%) after the sterilization process. Results are expressed as mean ± SD of the five sample replicates. Superscript in the same row indicates statistically significant differences between NaCl treatments (Tukey test, $p < 0.05$). Different capital letters in the same columns indicate statistically significant differences between olive presentation formats in each variety (olives, brine, and non-liquid olives) and NaCl treatment.

Varieties	Samples		Acrylamide (ng·g^{-1})	
			2 % NaCl	4 % NaCl
"Sevillana"	Olives	Stone	146 ± 9 [a C]	166 ± 9 [b C]
		Pitted	122 ± 10 [a B]	155 ± 9 [b B]
		Sliced	97 ± 7 [a A]	117 ± 7 [b A]
"Sevillana"	Brine	Stone	161 ± 7 [b C]	147 ± 11 [a C]
		Pitted	138 ± 11 [b B]	110 ± 9 [a B]
		Sliced	114 ± 8 [b A]	96 ± 8 [a A]
"Sevillana"	Non-liquid olives	Stone	426 ± 20 [ns C]	431 ± 13 [ns C]
		Pitted	343 ± 22 [ns B]	359 ± 20 [ns B]
		Sliced	231 ± 22 [ns A]	223 ± 20 [ns A]
"Hojiblanca"	Olives	Stone	111 ± 9 [a C]	137 ± 9 [b C]
		Pitted	91 ± 10 [a B]	115 ± 9 [b B]
		Sliced	73 ± 7 [a A]	97 ± 7 [b A]
"Hojiblanca"	Brine	Stone	120 ± 7 [b C]	107 ± 9 [a C]
		Pitted	101 ± 8 [b B]	90 ± 9 [a B]
		Sliced	85 ± 8 [b A]	73 ± 8 [a A]
"Hojiblanca"	Non-liquid olives	Stone	320 ± 20 [ns C]	331 ± 13 [ns C]
		Pitted	257 ± 22 [ns B]	269 ± 20 [ns B]
		Sliced	173 ± 21 [ns A]	163 ± 20 [ns A]
"Cacereña"	Olives	Stone	64 ± 5 [a C]	87 ± 6 [b C]
		Pitted	52 ± 5 [a B]	75 ± 6 [b B]
		Sliced	42 ± 6 [a A]	67 ± 6 [b A]
"Cacereña"	Brine	Stone	69 ± 7 [b C]	60 ± 7 [a B]
		Pitted	59 ± 7 [b B]	40 ± 6 [a AB]
		Sliced	49 ± 5 [b A]	41 ± 7 [a A]
"Cacereña"	Non-liquid olives	Stone	183 ± 20 [ns C]	171 ± 13 [ns C]
		Pitted	147 ± 22 [ns B]	159 ± 20 [ns B]
		Sliced	99 ± 22 [ns A]	103 ± 20 [ns A]

4. Conclusions

Californian-style olives are subjected to extreme temperatures during the sterilization process, promoting the formation of acrylamide. Significant amounts of acrylamide have been found in green ripe olives, although in lower concentrations than that in black ripe olives. The present study examined different mitigation strategies with the aim of reducing acrylamide formation in this foodstuff. According to the results obtained, it could be concluded that raw olives should be harvested at the green stage rather than allowing olives to reach later stages of maturity (green–yellow stage). This may mitigate sugar increases and the consequent formation of acrylamide after the sterilization process. Washing with water at 25 °C for at least 45 min but ideally for 2 h prior to lye treatment reduces the levels of acrylamide precursors in olives and, therefore, acrylamide formation during subsequent thermal treatment. Another point to control is the presentation format of olives. In this sense, stone olives have significantly more acrylamide content than sliced and pitted olives, since their smaller contact surface reduces the diffusion of acrylamide precursors from olives to the brine, which would enable greater acrylamide formation. Moreover, the addition of some additives, such as $CaCl_2$ and NaCl, commonly used in the elaboration of table olives to provide firmness and a salty taste to the fruit, should be controlled since their use increments acrylamide content in olives. These additives prevent diffusion of acrylamide precursors to the brine, thus increasing formation of the contaminant. This last consideration is not relevant when olives are canned without brine. Finally, due to the high presence of acrylamide in brine, it is recommended to wash olives prior to consumption. Results from this study are very useful for the table olive industry as they will enable it to identify critical points in the production of Californian-style green ripe olives and, in this way, control acrylamide formation in this foodstuff.

Author Contributions: Formal analysis, D.M.-V. and A.F.; investigation, D.M.-V.; supervision, D.M.-V., writing—original draft, D.M.-V., A.F., and E.M.-T.; writing—review and editing, D.M.-V., A.F., M.M. (Marta Mesías), M.M. (Manuel Martínez), M.D., and E.M.-T.; funding acquisition, D.M.-V. and M.M. (Manuel Martínez). All authors have read and agreed to the published version of the manuscript.

Funding: The research leading to these results received financial support from Junta de Extremadura (project IB18125) cofinanced by the European Regional Development Fund. Furthermore, this work has been also supported by research groups of Junta de Extremadura (GR18162).

Acknowledgments: The authors thank the Elemental and Molecular Analysis Service belonging to the Research Support Service of the University of Extremadura for development of the acrylamide determination method, and María Dolores López Soto and Elena Rodríguez Paniagua for their help in carried out this study. D.M.-V. is grateful to J.A. Pereira and A.M. Peres at the Centro de Investigação de Montanha (CIMO) (Polytechnic Institute of Bragança, Portugal) for the support in this research. We also thank the Research Institute of Agricultural Resources (INURA) for their help in the development of this work.

Conflicts of Interest: The authors state that they have no conflict of interest.

References

1. Hurtado, A.; Reguant, C.; Bordons, N.; Rozés, N. Lactic acid bacteria from fermented table olives. *Food Microbiol.* **2012**, *31*, 1–8. [CrossRef] [PubMed]

2. Campus, M.; Degirmencioglu, N.; Comunian, R. Technologies and Trends to improve table olive quality and safety. *Front. Microbiol.* **2018**, *9*, 617. [CrossRef]

3. IOOC (International Olive Oil Council). *Trade Standard Applying to Table Olives*; International Olive Oil Council COI/T20/Doc No 1; IOOC: Madrid, Spain, 2004.

4. Garrido Fernández, A.; Fernández Díez, M.J.; Adams, M.J. *Table Olives: Production and Processing*, 1st ed.; Chapman & Hall: London, UK, 1997.

5. Charoenprasert, S.; Mitchell, A. Influence of Californian-style black ripe olive processing on the formation of acrylamide. *J. Agric. Food Chem.* **2014**, *62*, 8716–8721. [CrossRef]

6. Casado, F.J.; Montaño, A. Influence of processing conditions on acrylamide content in black ripe olives. *J. Agric. Food Chem.* **2008**, *56*, 2021–2027. [CrossRef] [PubMed]

7. IARC (International Agency for Research on Cancer). Some industrial chemicals. In *IARC Monographs on the Evaluation for Carcinogenic Risk of Chemicals to Humans*; IARC: Lyon, France, 1994; Volume 60, pp. 435–453.

8. Casado, F.J.; Montaño, A.; Spitzner, D.; Carle, R. Investigations into acrylamide precursors in sterilized table olives: Evidence of a peptic fraction being responsible for acrylamide formation. *Food Chem.* **2013**, *141*, 1158–1165. [CrossRef] [PubMed]

9. EFSA (European Food Safety Authority). Scientific opinion on acrylamide in food. *EFSA J.* **2015**, *13*, 4104. Available online: https://efsa.onlinelibrary.wiley.com/doi/pdf/10.2903/j.efsa.2015.4104 (accessed on 10 June 2020).

10. Pérez-Nevado, F.; Cabrera-Bañegil, M.; Repilado, E.; Martillanes, S.; Martín-Vertedor, D. Effect of different baking treatments on the acrylamide formation and phenolic compounds in Californian-style black olives. *Food Control.* **2018**, *94*, 22–29. [CrossRef]

11. EC (European Commission). Commission Regulation (EU) 2017/2158 of 20 November 2017 establishing mitigation measures and benchmark levels for the reduction of the presence of acrylamide in food. *OJEU* **2017**, *304*, 24–44.

12. Tang, S.; Avena-Bustillos, R.J.; Lear, M.; Sedej, I.; Holstege, D.M.; Friedman, M.; McHugh, T.H.; Wang, S.C. Evaluation of thermal processing variables for reducing acrylamide in canned black ripe olives. *J. Food Eng.* **2016**, *191*, 124–130. [CrossRef]

13. Martín-Vertedor, D.; Fernández, A.; Hernández, A.; Arias-Calderón, R.; Delgado-Adámez, J.; Pérez-Nevado, F. Acrylamide reduction after phenols addition to Californian-style black olives. *Food Control* **2020**, *108*, 106888. [CrossRef]

14. Martín-Vertedor, D.; Rodrigues, N.; Marx, I.; Veloso, A.; Peres, A.M.; Pereira, J.A. Impact of thermal sterilization on the physicochemical-sensory characteristics of Californian-style black olives and its assessment using an electronic tongue. *Food Control* **2020**, *117*, 107369. [CrossRef]

15. López-López, A.; Beato, V.M.; Sánchez, A.H.; García-García, P.; Montaño, A. Effects of selected amino acids and water-soluble vitamins on acrylamide formation in a ripe olive model system. *J. Food Eng.* **2014**, *120*, 9–16. [CrossRef]

16. Montaño, A.; Casado, F.J.; Carle, R. Acrylamide in Table Olives. In *Acrylamide in Food. Analysis, Content and Potential Health Effects*; Gökmen, V., Ed.; Academic Press: London, UK, 2016; pp. 229–251.

17. Casado, F.J.; Sánchez, A.H.; Montaño, A. Reduction of acrylamide content of ripe olives by selected additives. *Food Chem.* **2010**, *119*, 161–166. [CrossRef]

18. Lodolini, E.M.; Cabrera-Bañegil, M.; Fernández, A.; Delgado-Adámez, J.; Ramírez, R.; Martín-Vertedor, D. Monitoring of acrylamide and phenolic compounds in table olive after high hydrostatic pressure and cooking treatments. *Food Chem.* **2019**, *286*, 250–259. [CrossRef] [PubMed]

19. Uceda, M.; Frías, L. Harvest dates. Evolution of the fruit oil content, oil composition and oil quality. In Proceedings of the II Seminario Oleícola Internacional, Córdoba, Spain, 6 October 1975; pp. 125–128.

20. Fernández, A.; Talaverano, M.I.; Pérez-Nevado, F.; Boselli, E.; Cordeiro, A.M.; Martillanes, S.; Foligni, R.; Martín-Vertedor, D. Evaluation of phenolics and acrylamide and their bioavailability in high hydrostatic pressure treated and fried table olives. *J. Food Process. Pres.* **2020**, *44*, e14384. [CrossRef]

21. Gonçalves, A.; Silva, E.; Brito, C.; Martins, S.; Pinto, L.; Dinis, T.L.; Luzio, A.; Martins-Gomes, C.; Fernandes-Silva, A.; Ribeiro, C.; et al. Olive tree physiology and chemical composition of fruits are modulated by different deficit irrigation strategies. *J. Sci. Food Agric.* **2020**, *100*, 682–694. [CrossRef] [PubMed]

22. Cabrera-Bañegil, M.; Pérez-Nevado, F.; Montaño, A.; Pleite, R.; Martín-Vertedor, D. The effect of olive fruit maturation in Spanish style fermentation with a controlled temperature. *LWT Food Sci. Technol.* **2018**, *91*, 40–47. [CrossRef]

23. Tsantili, E.; Christopoulos, M.V.; Pontikis, C.A.; Kaltsikes, P.; Kallianou, C.; Komaitis, M. Texture and other quality attributes in olives and leaf characteristics after preharvest calcium chloride sprays. *Hort. Sci.* **2008**, *43*, 1852–1856. [CrossRef]

24. Fadda, C.; Del Caro, A.; Sanguinetti, A.M.; Piga, A. Texture and antioxidant evolution of naturally green table olives as affected by different sodium chloride brine concentrations. *Int. J. Fat Oils* **2014**, *65*, e002.

Article

How Far Is the Spanish Snack Sector from Meeting the Acrylamide Regulation 2017/2158?

Marta Mesias *, Aouatif Nouali, Cristina Delgado-Andrade and Francisco J Morales

Institute of Food Science, Technology and Nutrition (ICTAN-CSIC), 28040 Madrid, Spain; anouali@ucm.es (A.N.); cdelgado@ictan.csic.es (C.D.-A.); fjmorales@ictan.csic.es (F.J.M.)

* Correspondence: mmesias@ictan.csic.es; Tel.: +34-91-549-2300; Fax: +34-91-549-3627

Received: 3 February 2020; Accepted: 21 February 2020; Published: 24 February 2020

Abstract: In 2017, the European Commission published Regulation 2017/2158 establishing mitigation measures and benchmark levels to reduce acrylamide in foods. Acrylamide was determined in seventy potato crisp samples commercialized in Spain. The aim was to update knowledge about the global situation in the snack sector and evaluate the effectiveness of mitigation strategies applied, especially since the publication of the Regulation. Results were compared with data previously published in 2004, 2008, and 2014, assessing the evolution over recent years. Average acrylamide content in 2019 (664 µg/kg, range 89–1930 µg/kg) was 55.3% lower compared to 2004, 10.3% lower compared to 2008 and practically similar to results from 2014. Results support the effectiveness of mitigation measures implemented by Spanish potato crisp manufacturers. However, 27% of samples exhibited concentrations above the benchmark level established in the Regulation (750 µg/kg), which suggests that efforts to reduce acrylamide formation in this sector must be continued. Besides the variability seen between samples, acrylamide significantly correlated with the color parameter a^*, which enables discrimination of whether potato crisps contain above or below benchmark content. The calculated margin of exposure for carcinogenicity was below the safety limit, which should be considered from a public health point of view.

Keywords: acrylamide; potato crisps; dietary intake; exposure; consumers

1. Introduction

Acrylamide is a chemical process contaminant formed when foods containing free asparagine and reducing sugars are cooked at temperatures above 120 °C in low moisture conditions [1]. It is identified by the International Agency for Research on Cancer as being probably carcinogenic to humans (group 2A) [2]. In 2015, the Expert Panel on Contaminants in the Food Chain (CONTAM) of the European Authority on Food Safety (EFSA) concluded that the presence of acrylamide in foods potentially increases the risk of developing cancer for consumers in all age groups [1]. Among the eleven food categories evaluated, potato fried products are generally the main contributor to total dietary acrylamide exposure [1].

There are several initiatives aiming to mitigate acrylamide formation in processed foods and its consequent acrylamide exposure and health concern. In the food processing sector, FoodDrinkEurope developed the so-called "Acrylamide Toolbox". This collects and updates the most effective technological strategies and recommendations in the food processing chain to mitigate the formation of acrylamide [3]. The Food and Drug Administration (FDA) also published guidance for acrylamide mitigation in the food industry to bring acrylamide levels down [4]. In parallel, the European Commission issued indicative values for the presence of acrylamide in foods in 2011 and 2013, based on EFSA monitoring data from 2007–2008 [5,6]. Subsequently, on 20 November 2017, the European Regulation (EU) 2017/2158 was published, establishing mitigation measures and benchmark levels for reducing the presence of acrylamide in food [7]. Although benchmark levels are not regulatory limits or safety thresholds, they serve to provide reference values for the industry. In the case that a product exceeds these values, the manufacturer should address the problem and apply relevant mitigation strategies in order to decrease acrylamide levels in the food. Focusing on potato-based products, including French fries and potato crisps, recommendations to mitigate the formation of this contaminant include selecting suitable potato varieties, controlling potato storage and transport, monitoring recipe, and process conditions, and providing information to end users about adequate cooking practices. For potato crisps, the first indicative value for acrylamide was set at 1000 µg/kg [5,6], though subsequent revisions have reduced this benchmark level to 750 µg/kg [7].

Control of any step in the manufacturing process of fried potatoes is especially important to reduce exposure to acrylamide. Variations in total dietary exposure to this contaminant of up to 80% can be expected in fried potatoes, depending on the conditions of potato frying [1]. As a result of the commitment of the potato snack sector to follow indications stipulated in the "Acrylamide Toolbox", various studies have reported the successful application of acrylamide mitigation strategies in potato snacks over recent years. Powers et al. [8] described a statistically significant downward trend in mean acrylamide levels in potato crisps in Europe from 763 µg/kg in 2002 to 358 µg/kg in 2011 (a decrease of 53%). The same authors published an updated evaluation in 2016, describing mean concentration to be 412 µg/kg [9]. While these data represent a 46% reduction since 2002, a leveling off has been observed since 2011 [9]. In a previous study, our research group observed a decrease in acrylamide levels in potato crisps in the Spanish market from 1484 µg/kg in 2004 to 629 µg/kg in 2014, representing an overall reduction of 57.6% [10]. Despite this decline, 17.5% of the samples in 2014 registered values higher than the indicative value recommended by the European Commission for potato crisps, at the time of publication. Parallel to this, our research team, together with some others, has established the existence of a direct correlation between the CIELab color parameter *a** and acrylamide content in this foodstuff, proposing *a** as a useful predictor of the acrylamide formation in fried potatoes [11–15].

In the present investigation, the acrylamide content of potato crisps marketed in Spain was re-evaluated one year after introduction of the Regulation with a view to investigate the current state of the snack food sector and degree of compliance with the Regulation. Results were considered alongside results produced during the period 2004 to 2014. Additionally, suitability of the CIElab parameter *a** for predicting acrylamide levels in potato crisps was explored.

2. Material and Methods

2.1. Chemicals and Materials

Acrylamide standard (99%), potassium hexacyanoferrate (II) trihydrate (98%, Carrez-I), and zinc acetate dehydrate (>99%, Carrez-II) were obtained from Sigma (St. Louis, MO, USA). $^{13}C_3$-labelled acrylamide (99% isotopic purity) was obtained from Cambridge Isotope Laboratories (Andover, MA, USA). Formic acid (98%), methanol (99.5%) and hexane were from Panreac (Barcelona, Spain). Deionized water was obtained from a Milli-Q Integral 5 water purification system (Millipore, Billerica, MA, USA). All other chemicals, solvents and reagents were of analytical grade. Reversed-phase Oasis-HLB cartridges (30 mg, 1 mL) were from Waters (Mildford, MA, USA). Syringe filter units (0.45 µm and 0.20 µm, cellulose) were purchased from Análisis Vínicos (Tomelloso, Ciudad Real, Spain).

2.2. Samples

Seventy commercial potato crisps from 33 different producers were purchased in several Spanish supermarkets in February 2019. Of the producers, 11 supplied the present study with more than one brand, whilst the remaining 22 supplied a single brand only. Samples containing different flavorings and/or added spices were excluded from sampling in order to avoid bias during data interpretation. Potato crisps were classified according to producer, type of frying oil (sunflower oil, olive oil, and other/undefined oil), type of cut (smooth, wavy), oily appearance (oily, semi-oily, non-oily), and type of container (light protected, partly-light protected, non-light protected). Potato crisps (100–280 g, according to the size of the bag) were mixed and ground to ensure homogeneous distribution of potential hotspots. A portion (ca. 80 g) was distributed into two containers and stored under vacuum and light protected conditions at 4 °C, until analysis.

2.3. Determination of CIELAB Color

Color measurements were made at room temperature using a HunterLab Spectrophotometer 150 CM-3500D colorimeter (Hunter Associates laboratory, Stamford, CT, USA). Three independent measurements of a^* (redness), b^* (yellowness), and L^* (lightness) parameters were carried out on different areas of the ground potato crisps in order to consider the non-homogeneous distribution of color within the same batch of the product. E index was calculated according to the following equation: $E = (L^2 + a^2 + b^2)^{1/2}$. Equipment was calibrated with a standard calibration white plate CR-A43 (L^*/93.80, a^*/0.3156, b^*/0.3319).

2.4. LC-ESI-MS-MS Determination of Acrylamide

Acrylamide was determined in potato crisp samples as described by Mesias and Morales [10] and based on the ISO:EN:16618:2015 method. The recovery rate of acrylamide spiked to the samples was between 90 and 106%. Relative standard deviations (RSD) for precision, repeatability, and reproducibility of analyses were calculated as 2.8%, 1.2% and 2.5%, respectively. The limit of the quantitation was set at 20 µg/kg, complying with performance criteria set by EU Regulation 2017/2158 [7]. Accuracy of the results was demonstrated for potato crisps and pre-cooked French fries in four proficiency tests launched by the Food Analysis Performance Assessment Scheme (FAPAS) program, yielding a *z*-score of –0.2 (Test 3071, Feb–March 2017), –0.3 (Test 3080, Feb–March 2018), 0.0 (Test 3085, Sep–Oct 2018), and 0.3 (Test 3089, Feb-2019). Results of acrylamide were expressed as µg/kg of sample. Analysis was done in duplicate.

2.5. Dietary Exposure Assessment

Categories of dietary exposure to acrylamide from potato crisps were estimated by combining data for total per capita consumption of potato crisps (1.34 kg/person/year) established by the Spanish Ministry of Agriculture, Food and Environment [16], and acrylamide content of the samples. An average body weight (bw) of 70 kg was used to estimate total daily intake of acrylamide from potato crisps for the total population and expressed as µg/kg bw/day.

The margin of exposure (MOE) approach was applied to carry out risk assessment for acrylamide provided by potato crisps. MOE was calculated as the BMDL value divided by the respective total acrylamide intake, considering 430 µg/kg bw/day as the $BMDL_{10}$ value for neurotoxicity (peripheral nerve axonal degeneration in male rats) and 170 µg/kg bw/day for carcinogenicity (harderian gland adenocarcinomas in mice). This is dictated in the EFSA opinion report on acrylamide [1].

2.6. Statistical Analysis

Statistical analyses were performed using SPSS version 23.0 (SPSS Inc., Chicago, IL, USA). Data were expressed as mean ± standard deviation (SD). Student *t*-test and analysis of variance (ANOVA one-way) followed by the Fisher's test were used to identify the overall significance of differences between variables. Homogeneity of variances was determined using the Levene test. Relationships between the different variables were evaluated by computing Spearman's linear correlation coefficients. The significance of all statistical parameters was evaluated at the level of $p < 0.05$.

3. Results and Discussion

3.1. Nutritional Composition of Potato Crisps

Table S1 depicts the average nutritional composition of potato crisps as declared by the manufacturer. Energy values (439–589 kcal/100 g) were close for all samples, with higher differences being observed in relation to other parameters. Total fat content ranged between 13.2 and 40.7 g/100 g, with saturated fat being between 1.4 and 6.1 g/100 g. Carbohydrates exhibited a mean value of 50.1 g/100 g (38.0–72.1 g/100 g) and sugar content was 0.6 g/100 g (0.1–4.7 g/100 g). Fiber showed values from 0.5 to 7.7 g/100 g, and protein values ranged from 1.0 to 7.8 (maximum) g/100 g. The maximum salt level was 1.7 g/100 g, although 3.0% of samples claimed to be 'no added-salt'.

3.2. Acrylamide Levels in Potato Crisps

Seventy (non-flavored) classical potato crisps commercialized in Spain coming from 33 different snack producers were collected during February 2019 and analyzed for acrylamide. All samples showed an acrylamide content greater than the LOQ, ranging from 89 to 1930 µg/kg (Table 1). Mean, median, and 95th percentile were 664 µg/kg, 569 µg/kg, and 1576 µg/kg, respectively. These values were higher than those reported by the EFSA for the category of potato crisps made from fresh potatoes (mean value: 392 µg/kg, 95th percentile: 949 µg/kg, n = 31467) in European countries [1]. However, results were in line with information provided by other authors in recent years. Hai et al. [17] found levels ranging from 25 to 1620 µg/kg in potato chips from Hanoi (Vietnam), whilst Kafouris et al. [18] described an average value of 642 µg/kg in potato crisps from Cyprus (10–2193 µg/kg). Higher concentrations have been reported in potato crisps from Italy (173–3444 µg/kg, average value: 1162 µg/kg) [19] and lower value (mean value 475 µg/kg) have been described for potato crisps collected from several Portuguese local markets [20].

Table 1. Statistical analysis of the acrylamide levels (µg/kg) of different potato crisp categories.

Category	Mean ± SD	Median	Minimum	Maximum	95th	n
Total	664 ± 387	569	89	1930	1576	70
Type of oil						
Sunflower	689 ± 387a	627	199	1930	1271	41
Olive oil	624 ± 333a	560	89	1381	1280	20
Other or undefined oil	642 ± 524a	459	243	1815	1563	9
Type of cut						
Smooth	642 ± 369a	561	89	1889	1286	59
Wavy	786 ± 473a	577	315	1930	1548	11
Appearance						
Oily	614 ± 287a	541	241	1275	1212	35
Semi-oily	886 ± 592b	656	221	1930	1102	12
Non-oily	626 ± 368ab	561	89	1815	1838	23
Light protection						
Light protected	695 ± 410a	599	199	1930	1650	47
Partly-light protected	587 ± 332a	539	89	1381	750	18
Non-light protected	653 ± 385a	541	259	1271	1349	5

Different letters in the same column mean significant differences within the same category ($p < 0.05$). SD: standard deviation. 95th: 95 percentile.

Figure 1 shows acrylamide distribution in the seventy measured samples of potato crisps. Data exhibited high variability in acrylamide content between samples, in accordance with previously published results [10]. Four outliers were identified in the global distribution. As explained by Mesias and Morales [10], the 90th percentile is identified as the signal value of the dataset. Samples found with values higher than this should be especially evaluated in order to mitigate acrylamide formation.

Nevertheless, nineteen of the potato crisps analyzed exceeded benchmark level for acrylamide as established by the regulation for this foodstuff (750 µg/kg) [7].

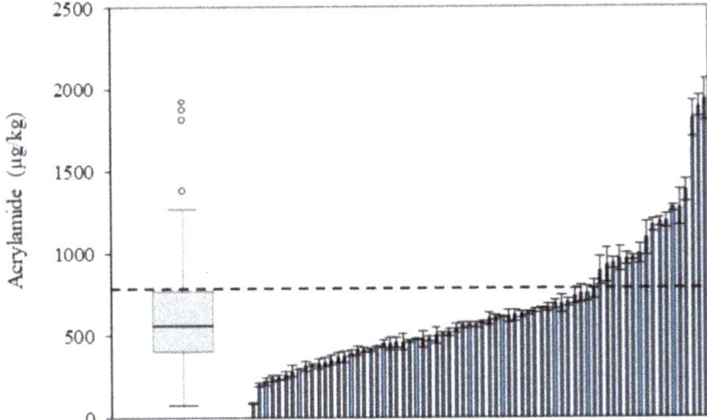

Figure 1. Box and whisker plot and distribution graph for the acrylamide content in Spanish potato crisps. Symbols: o outliers. The dotted line indicates benchmark level (750 µg/kg) established by the European Commission [7].

Seasonality of the potato tuber is a critical factor in strategies for acrylamide reduction since it has a great influence on levels of acrylamide precursors. Powers et al. [8] reported levels ranging from 528 µg/kg in potato crisps collected in the first six months of the year to 372 µg/kg in those collected during the second six months. Additionally, acrylamide precursor levels will depend on post-harvest storage conditions due to inadequate temperature control—for instance enabling temperatures to fall below 10 °C will increase the degradation of starch, whilst increasing the content of reducing sugars [21,22], a part of the potato variety and tuber storage time [23]. In the present study, potato crisp samples were collected in February. This is the most unfavorable period for potato producers in Spain as it is very likely that the raw material comes from stored potato. This contains higher levels of reducing sugars, thus leading to higher levels of acrylamide during frying.

Besides the nutritional composition, additional information was considered in order to evaluate the possible relationship between different processing conditions for potatoes and their acrylamide content. Thus, samples were grouped according to the type of frying oil (sunflower oil, olive oil, and other or undefined oil), type of cut (smooth and wavy), oily appearance (oily, semi-oily, and non-oily), and type of container (light- protected, partly-light protected, and non-light protected) (Table 1). Acrylamide levels were not directly affected by the type of frying oil or with the extent to which the container was light protected. Regarding the type of cut, wavy samples displayed a higher mean acrylamide concentration (786 µg/kg, $n = 11$) than smooth potato crisps (642 µg/kg, $n = 59$). These findings were not significant due to sample variability and potentially due to the unequal number of samples found for each category. Although the type of cut involves different dimensions and a potentially different acrylamide content in potato crisps, Powers et al. [9] did not observe significant differences in the acrylamide content of thick and ridge/wave potato crisps in a European sample. In our previous study, the accumulation of visually detected oil drops on the crisp's surface was related to higher acrylamide levels [10]. In the present study, although significant differences ($p < 0.05$) were again observed with respect to the presence of oil drops on the container (Table 1), it was not possible to establish any relationship with this data.

Samples were also grouped according to the producer (Figure 2). Acrylamide content varied greatly between the 33 producers. Mean acrylamide content ranged from 89 µg/kg to 1561 µg/kg between different producers. Eight of the producers had mean values exceeding the acrylamide benchmark value set by the European regulation, especially producer 23 (1561 µg/kg) and 30 (1055 µg/kg). Nearly one-third of the producers provided more than one brand for this study. Large acrylamide variations were also observed within the same producer; for instance, samples from producer 30 ranged from 295 µg/kg to 1815 µg/kg. High variability in acrylamide content of potato crisps suggests differences in the raw material (potato variety and post-harvest storage) and processing conditions applied. Although this information is not available, it is assumed that potato crisp producers marketing a unique brand are small companies with a nearby geographical distribution of their product, in contrast to large companies marketing several brands throughout the country. From a managerial point of view, large snacking companies would have more resources to implement mitigation strategies; for instance, they are more likely to be able to select the most appropriate potato tuber. However, only four of the 22 small potato crisp producers exceeded the benchmark value for acrylamide.

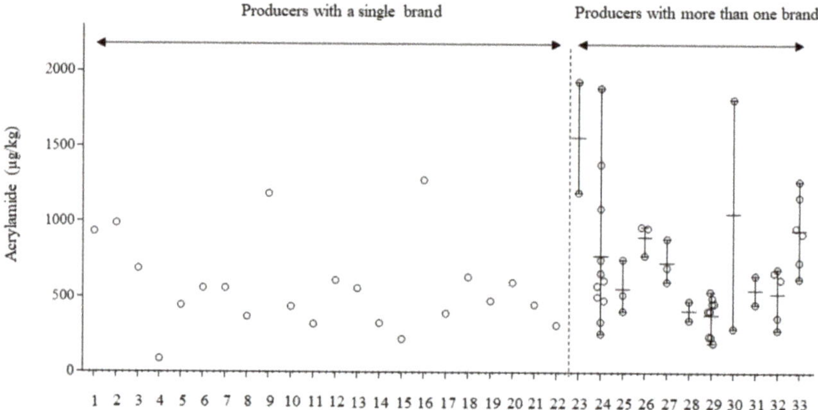

Figure 2. Acrylamide content in potato crisp samples from the Spanish market, grouped according to their producers ($n = 33$). Among them: producers with a single brand ($n = 22$) and producers with more than one brand ($n = 11$). Circles represent individual values; horizontal lines represent mean values and vertical lines represent the range for each producer.

3.3. Acrylamide–Color Relationship in Potato Crisps

One of the critical factors to establish the end-point of frying is the color of potato chips [24]. Samples in the present study were characterized according to the CIELab color scale. Mean ± standard deviation, minimum, and maximum values for a^*, b^*, L^*, and E are presented in Table 2. Again, a huge variability was observed between the samples, with values ranging from −0.62 to 5.53 for a^*, from 16.08 to 29.22 for b^*, from 50.04 to 68.01 for L^*, and from 53.74 to 72.24 for E. A significant correlation was found between acrylamide and L^* ($r^2 = -0.2889$, $p = 0.0153$) and E ($r^2 = -0.2511$, $p = 0.0360$), indicating that browning and lower luminosity in potato chips led to greater formation of this contaminant. The correlation with b^* was not found to be significant; however, a strong relationship was displayed with a^* ($r^2 = 0.6736$, $p < 0.0001$) (Figure 3A). Two essential factors would explain that the curve does not go through zero. On the one hand, the color parameter a^* in the fried potato has not been corrected by the color in the raw potato. On the other hand, the contribution to the color of the frying oil should be also considered since nearly 30% of the final product is oil that affects the color. However, even considering this limitation, the relationship is strongly significant. Previous studies have already reported a direct correlation between a^* and acrylamide in this foodstuff; thus, a^* has been proposed as a useful predictor of acrylamide formation in fried potatoes [11–15]. Besides the high variability according to sample type, producers, and acrylamide level, our results pointed out that the a^* parameter is capable of discriminating ($p < 0.0001$) between samples that contain above and below benchmark level of acrylamide (750 µg/kg) (Figure 3B).

Table 2. CIELab parameters for potato crisps.

CIELab Parameter	Mean ± SD	Minimum	Maximum
a^*	2.64 ± 0.13	−0.62	5.53
b^*	24.75 ± 0.24	16.08	29.22
L^*	56.97 ± 0.27	50.04	68.01
E	62.21 ± 0.38	53.74	72.24

SD: standard deviation

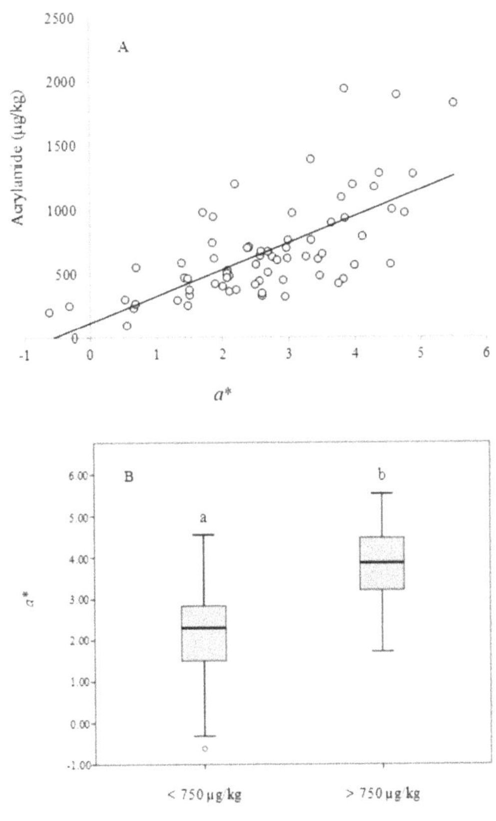

Figure 3. Relationship between color and acrylamide content (µg/kg) in potato crisps marketed in Spain. Linear regression between acrylamide content and the color parameter a^* ($r^2 = 0.6736$, $p < 0.0001$) (**A**) and box-and-whisker plot of color parameter a^* according to the acrylamide levels using the benchmark level as the threshold (750 µg/kg) (**B**).

3.4. Evaluation of Acrylamide Exposure from Spanish Commercialized Potato Crisps

Exposure to acrylamide from potato crisps in the Spanish population was estimated considering total data per capita consumption of this foodstuff as indicated by the Spanish Ministry of Agriculture, Food and Environment (1.34 kg/person/year) [16], and acrylamide content in present samples. Considering the minimum and maximum acrylamide content in the overall sample, exposure to this contaminant was seen to range between 0.33 and 7.09 µg/person/day, with a mean value of 2.44 µg/person/day. Assuming an average body weight (bw) of 70 kg, mean daily intake of acrylamide from potato crisps for the total population was estimated to be 0.035 µg/kg bw/day, ranging from 0.005 to 0.101 µg/kg bw/day. Median values matched with previous estimations made in 2015, in which average contribution of potato crisps to dietary acrylamide exposure in Spain was also calculated to be 0.035 µg/kg (range: 0.006–0.12 µg/kg bw/day) [10].

Acrylamide exposure estimated in the present assay was similar to that observed for other countries in studies with different population sectors (Table 3). The average value was similar to the median reported in adults from Italy [19], higher than mean exposure in adults from Belgium [25] and much lower than in the general population from Denmark [26]. Adolescents generally present greater exposure [18,19,25], except those from Poland [27], with median values being close to the mean described in the present study. In contrast, elderly and very elderly populations have shown lower acrylamide exposure due to lower consumption of potato crisps [19].

Table 3. Dietary acrylamide exposure and contribution of fried potatoes in different studies, countries, and population groups.

Country	Total Intake (µg/kg bw/day)	Potato Chips/Crisps (%)	Partial Contribution (µg/kg bw/day)		Age (years)	Reference
			Median[2] or Mean[3]	95th		
Italy	n.a.[1]	n.a.[1]	0.070[2]	0.387	Toddlers	[19]
			0.112[2]	0.238	Other children	
			0.075[2]	0.201	Adolescents	
			0.030[2]	0.097	Adults	
			0.004[2]	0.029	Elderly	
			0.001[2]	0.019	Very elderly	
Poland	0.09	50	0.045[2]	0.500	Teenager girls	[27]
	0.13	34	0.044[2]	0.570	Teenager boys	
Denmark	0.31	46[4]	0.143[3]	0.382	General population	[26]
Cyprus	0.8	14	0.112[3]	0.252	Adolescents	[18]
Belgium	0.72	5.9	0.042[3]	0.143	Children	[15]
	0.48	18.8	0.090[3]	0.297	Adolescents	
	0.33	5.8	0.019[3]	0.063	Adults	

[1] n.a.: not available. [2] Median value. [3] Mean value. [4] Fried potatoes including French fries

Risk characterization for acrylamide in potato crisps was conducted taking MOE values of 125 and 10,000 as values indicating no concern for neurotoxicity and carcinogenicity in people, respectively [1]. As mentioned above, MOE was calculated as the BMDL value divided by respective total acrylamide intake; 430 µg/kg bw/day was considered as the $BMDL_{10}$ value for neurotoxicity, and 170 µg/kg bw/day as the value for carcinogenicity, as dictated in EFSA opinion on acrylamide [1]. In this respect, mean MOE value of 12,348 was obtained for neurotoxicity (range: 92,122–4248 for minimum and maximum exposure). Even the maximum value was above the safety limit of 125, which indicates no health concern. In contrast, when comparing with the $BMDL_{10}$ for carcinogenicity, a mean value of 4882 was calculated (range: 36,420–1679 for minimum and maximum exposure). In this case, MOE was below the safety limit of 10,000, suggesting that it should be considered from a public health point of view. This remarkable outcome is in line with data reported in several studies [1,19,27].

3.5. Evolution of Acrylamide Levels in Spanish Commercialized Potato Crisps in Recent Years

Acrylamide content in commercialized potato chips analyzed in this study was compared with results previously obtained by the same research group in 2014 [10], 2008 [28], and 2004 [29]. Mesias and Morales [10] showed a clear downward trend in acrylamide content of potato crisps marketed in Spain over the last 10 years. Mean values decreased from 1484 µg/kg in 2004 to 629 µg/kg in 2014, representing an overall reduction of 57.6%. Analyzing the whole period monitored between 2004 and 2019, the overall reduction was 55.2%. Statistical analysis revealed a significant reduction in the period 2004–2008 (Figure 4); however, this has since been followed by a period of leveling off, with no significant differences between mean acrylamide levels for 2008 (740 µg/kg), 2014 (629 µg/kg), and 2019 (664 µg/kg) being seen. Similar observations were displayed in the 90th and 95th percentiles. Downward trends were also seen in these values from 2004 to 2019 (Figure 4). 90th percentile values, considered as signal values, decreased from 2270 µg/kg in 2004 to 1377 µg/kg in 2008, and 1136 µg/kg in 2014, slightly increasing back up to 1187 µg/kg in 2019. Similarly, 95th percentile values decreased from 3805 µg/kg in 2004 to 1981 µg/kg in 2008, and 1233 µg/kg in 2014, increasing again to 1576 µg/kg in 2019. These findings are in agreement with results reported by Powers et al. [9]. These authors evaluated the trend in acrylamide concentration in potato crisps in Europe from 2002 to 2016, describing a 46% reduction overall throughout the period. Mean values decreased from 763 µg/kg in 2002 to 412 µg/kg in 2016. However, as in the present study, there were significant reductions between 2002 and 2011, but, since then, a leveling off has been observed, with means being even slightly greater in recent years. Thus, the year when the lowest content of acrylamide was achieved was 2011. In our case, lowest concentrations were observed in 2014.

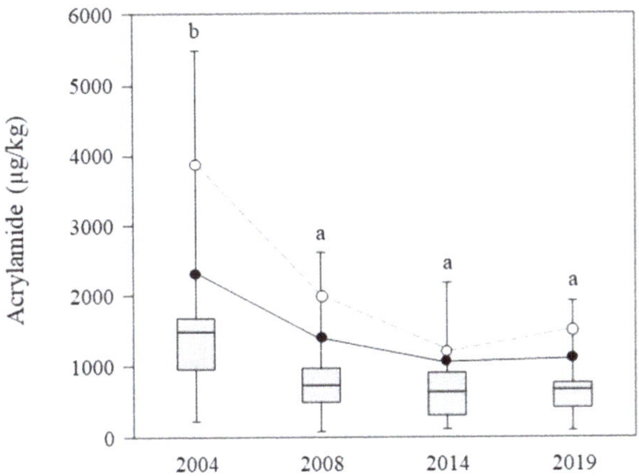

Figure 4. Temporal patterns for acrylamide content (µg/kg) in Spanish potato crisps sampled from 2004 to 2019. ● represents 90th percentile, ○ represents 95th percentile.

Similarly, Claeys et al. [25] also reported a significant decrease in mean acrylamide levels in potato crisps from the Belgian market, when comparing the periods 2002–2007 and 2008–2013 (average reduction 38%). In this case, concentrations, 95th percentile values, and maximum levels decreased from 609 µg/kg (mean), 1500 µg/kg (95th), and 3200 µg/kg (maximum) in the period 2002–2007 to 375 µg/kg (mean), 725 µg/kg (95th) and 1300 µg/kg (maximum), respectively, in the period 2008–2013.

As mentioned before, 17.5% of samples taken in 2015 registered higher values than those recommended by the European Commission that year (1000 µg/kg), whilst, in the present study, 27% of analyzed potato crisps exceeded the updated benchmark value (750 µg/kg). Results illustrate the need to reduce levels of acrylamide in potato crisps since concentrations have not yet been adapted to the new recommendations. In recent years, acrylamide reduction measures in potato-based products in the acrylamide toolbox have been proposed and updated [3], as has industry guidance regarding acrylamide in foods reported by the Food and Drug Administration [4]. In addition, risk management measures for acrylamide were included and revised in the European Regulation [7].

4. Conclusions

Although it has been demonstrated that mitigation strategies have been successfully applied to Spanish industrial potato crisps in recent years, the greatest reduction in acrylamide levels was achieved during years close to the identification of acrylamide in foods as a health concern. Following this, declining trends have ceased and nearly one-third of samples fail to meet benchmark value established by the current European Regulation. Thus, potato crisp manufacturers must continue efforts to reduce acrylamide formation in this foodstuff in order to reach levels as low as reasonably achievable, whilst also considering the potentially pending downward revision of the benchmark level. This new study corroborates the suitability of the color parameter a^* for helping producers to control the quality of potato crisps introduced into the market, so that they meet quality standards indicated by the European Regulation concerning acrylamide levels.

Supplementary Materials: The following are available online at http://www.mdpi.com/2304-8158/9/2/247/s1, Table S1. Average nutritional composition of the whole dataset of potato crisps as provided by the manufacturer. Data are expressed as mean ± standard deviation (SD) per 100 g of sample

Author Contributions: Conceptualization, M.M., C.D.-A., and F.J.M.; Methodology, A.N.; Software, M.M.; Validation, M.M., C.D.-A., and F.J.M.; Formal Analysis, A.N.; Investigation, M.M., C.D.-A., and F.J.M.; Resources, F.J.M.; Data Curation, M.M., C.D.-A., and F.J.M.; Writing—Original Draft Preparation, M.M.; Writing—Review and Editing, M.M., C.D.-A., and F.J.M.; Supervision, M.M., C.D.-A., and F.J.M.; Funding Acquisition, M.M., C.D.-A., and F.J.M. All authors have read and agreed to the published version of the manuscript.

Funding: This research was partly funded by the Ministry of Science, Innovation and Universities (Spain): project ACRINTAKE (RTI2018-094402-B-I00, MCIU/AEI/FEDER, UE), Comunidad of Madrid, and European funding from FSE and FEDER programs: Project S2018/BAA-4393, AVANSECAL-II-CM. The authors thank Inmaculada Alvarez and Beatriz Díaz for their technical assistance.

Acknowledgments: This work was partially supported by the Ministry of Science, Innovation, and Universities (Spain) under the project ACRINTAKE (RTI2018-094402-B-I00, MCIU/AEI/FEDER, UE) and by the Community of Madrid, and European funding from FSE and FEDER programs under the project S2018/BAA-4393, AVANSECAL-II-CM.

Conflicts of Interest: The authors declare that they have no conflict of interest.

References

1. EFSA (European Food Safety Authority). Scientific opinion on acrylamide in food. *EFSA J.* **2015**, *13*, 4104. Available online: https://efsa.onlinelibrary.wiley.com/doi/pdf/10.2903/j.efsa.2015.4104 (accessed on 12 December 2019).

2. IARC (International Agency for Research on Cancer). Some industrial chemicals. In *IARC Monographs on the Evaluation for Carcinogenic Risk of Chemicals to Humans*; IARC: Lyon, France, 1994; Volume 60, pp. 435–453.

3. FDE (FoodDrinkEurope). 2019. Available online: https://www.fooddrinkeurope.eu/uploads/publications_documents/FoodDrinkEurope_Acrylamide_Toolbox_2019.pdf (accessed on 12 December 2019).

4. FDA (Food and Drug Administration). Guidance for Industry Acrylamide in Foods. 2016. Available online: https://www.fda.gov/media/87150/download (accessed on 5 December 2019).

5. Commission Recommendation. Investigations into the levels of acrylamide in food. *OJEU* **2011**. Available online: http://ec.europa.eu/food/food/chemicalsafety/contaminants/recommendation_10012011_acrylamide_food_en.pdf. (accessed on 1 December 2019).

6. Commission Recommendation. Investigations into the levels of acrylamide in food. *OJEU* **2013**, *L301/15*.

7. EC (European Commission). Commission Regulation (EU) 2017/2158 of 20 November 2017 establishing mitigation measures and benchmark levels for the reduction of the presence of acrylamide in food. *OJEU* **2017**, *L304*, 24–44.

8. Powers, S.J.; Mottram, D.S.; Curtis, A.; Halford, N.G. Acrylamide concentrations in potato crisps in Europe from 2002 to 2011. *Food Addit. Contam. Part A* **2013**, *30*, 1493–1500. [CrossRef]

9. Powers, S.J.; Mottram, D.S.; Curtis, A.; Halford, N.G. Acrylamide levels in potato crisps in Europe from 2002 to 2016. *Food Addit. Contam. Part A* **2017**, *34*, 2085–2100. [CrossRef]

10. Mesias, M.; Morales, F.J. Acrylamide in commercial potato crisps from Spanish market: Trends from 2004 to 2014 and assessment of the dietary exposure. *Food Chem. Toxicol.* **2015**, *81*, 104–110. [CrossRef]

11. Gökmen, V.; Senyuva, H.Z.; Dülek, B.; Çetin, A.E. Computer vision-based image analysis for the estimation of acrylamide concentrations of potato chips and french fries. *Food Chem.* **2007**, *101*, 791–798. [CrossRef]

12. Behrens, T.K.; Liebert, M.L.; Peterson, H.J.; Smith, H.J.; Sutliffe, J.T.; Day, A.; Mack, J. Changes in School Food Preparation Methods Result in Healthier Cafeteria Lunches in Elementary Schools. *Am. J. Prev. Med.* **2018**, *54*, S139–S144. [CrossRef]

13. Mesias, M.; Delgado-Andrade, C.; Holgado, F.; Morales, F.J. Acrylamide content in French fries prepared in households: A pilot study in Spanish homes. *Food Chem.* **2018**, *260*, 44–52. [CrossRef]

14. Mesias, M.; Delgado-Andrade, C.; Holgado, F.; Morales, F.J. Acrylamide content in French fries prepared in food service establishments. *LWT Food Sci. Technol.* **2019**, *100*, 83–91. [CrossRef]

15. Mesias, M.; Delgado-Andrade, C.; Holgado, F.; Morales, F.J. Acrylamide in French fries prepared at primary school canteens. *Food Funct.* **2020**, (in press). [CrossRef] [PubMed]

16. MAPA (Ministry of Agriculture, Fisheries and Foods). Datos de consumo alimentario. Informe del consume alimentario en España. 2018. Available online: https://www.mapa.gob.es/es/alimentacion/temas/consumo-y-comercializacion-y-distribucion-alimentaria/20190807_informedeconsumo2018pdf_tcm30-512256.pdf (accessed on 10 December 2019).

17. Hai, Y.D.; Tran-Lam, T.T.; Nguyen, T.Q.; Vu, N.D.; Ma, K.H.; Le, G.T. Acrylamide in daily food in the metropolitan área of Hanoi, Vietnam. *Food Addit. Contam. Part B* **2019**, *12*, 159–166. [CrossRef] [PubMed]

18. Kafouris, D.; Stavroulakis, G.; Christofidou, M.; Iakovou, X.; Christou, E.; Paikousis, L.; Christodoulidou, M.; Ioannou-Kakouri, E.; Yiannopoulos, E. Determination of acrylamide in food using a UPLC–MS/MS method: Results of the official control and dietary exposure assessment in Cyprus. *Food Addit. Contam. Part A* **2018**, *35*, 1928–1939. [CrossRef] [PubMed]

19. Esposito, F.; Nardone, A.; Fasano, E.; Triassi, M.; Cirillo, T. Determination of acrylamide levels in potato crisps and other snacks and exposure risk assessment through a Margin of Exposure approach. *Food Chem. Toxicol.* **2017**, *108*, 249–256. [CrossRef]

20. Molina-Garcia, L.; Santos, C.S.P.; Melo, A.; Fernandes, J.O.; Cunha, S.C.; Casal, S. Acrylamide in chips and French fries: A novel and simple method using xanthydrol for its GC-MS determination. *Food Anal. Methods* **2015**, *8*, 1436–1445. [CrossRef]

21. De Wilde, T.; De Meulenaer, B.; Mestdagh, F.; Govaert, Y.; Vandeburie, S.; Ooghe, W.; Fraselle, S.; Demeulemeester, K.; Van Peteghem, C.; Calus, A.; et al. Influence of Storage Practices on Acrylamide Formation during Potato Frying. *J. Agric. Food Chem.* **2005**, *53*, 6550–6557. [CrossRef]

22. Muttucumaru, N.; Powers, S.J.; Elmore, J.S.; Briddon, A.; Mottram, D.S.; Halford, N.G. Evidence for the complex relationship between free amino acid and sugar concentrations and acrylamide-forming potential in potato. *Ann. Appl. Biol.* **2014**, *164*, 286–300. [CrossRef]

23. Elmore, J.S.; Briddon, A.; Dodson, A.T.; Muttucumaru, N.; Halford, N.G.; Mottram, D.S. Acrylamide in potato crisps prepared from 20 UK-grown varieties: Effects of variety and tuber storage time. *Food Chem.* **2015**, *182*, 1–8. [CrossRef]

24. Pedreschi, F.; Mery, D.; Marique, T. Quality evaluation and control of potato chips and French fries. In *Computer Vision Technology for Food Quality Evaluation*; Sun, D., Ed.; Academic Press: Cambridge, MA, USA, 2008; pp. 545–566.

25. Claeys, W.; De Meulenaer, B.; Huyghebaert, A.; Scippo, M.L.; Hoet, P.; Matthys, C. Reassessment of the acrylamide risk: Belgium as a case-study. *Food Control* **2016**, *59*, 628–635. [CrossRef]

26. Jakobsen, L.S.; Granby, K.; Knudsen, V.K.; Nauta, M.; Pires, S.M.; Poulsen, M. Burden of disease of dietary exposure to Acrylamide in Denmark. *Food Chem. Toxicol.* **2016**, *90*, 151–159. [CrossRef]

27. Wyka, J.; Tajner-Czopek, A.; Broniecka, A.; Piotrowska, E.; Bronkowska, M.; Biernat, J. Estimation of dietary exposure to Acrylamide of Polish teenagers from an urban environment. *Food Chem. Toxicol.* **2015**, *75*, 151–155. [CrossRef]

28. Arribas-Lorenzo, G.; Morales, F.J. Dietary exposure to acrylamide from potato crisps to the Spanish population. *Food Addit. Contam.* **2009**, *26*, 289–297. [CrossRef] [PubMed]
29. Rufián-Henares, J.A.; Morales, F.J. Determination of acrylamide in potato chips by a reversed-phase LC-MS method based on the stable isotope dilution assay. *Food Chem.* **2006**, *97*, 555–562. [CrossRef]

 *foods*

Article

Are Household Potato Frying Habits Suitable for Preventing Acrylamide Exposure?

Marta Mesias *, Cristina Delgado-Andrade and Francisco J. Morales

Institute of Food Science, Technology and Nutrition, ICTAN-CSIC, José Antonio Novais 10, 28040 Madrid, Spain; cdelgado@ictan.csic.es (C.D.-A.); fjmorales@ictan.csic.es (F.J.M.)
* Correspondence: mmesias@ictan.csic.es

Received: 22 May 2020; Accepted: 12 June 2020; Published: 17 June 2020

Abstract: A survey was conducted of 730 Spanish households to identify culinary practices which might influence acrylamide formation during the domestic preparation of french fries and their compliance with the acrylamide mitigation strategies described in the 2017/2158 Regulation. Spanish household practices conformed with the majority of recommendations for the selection, storing and handling of potatoes, with the exception of soaking potato strips. Olive oil was the preferred frying oil (78.7%) and frying pans were the most common kitchen utensils used for frying (79.0%), leading to a higher oil replacement rate than with a deep-fryer. Although frying temperature was usually controlled (81.0%), participants were unaware of the maximum temperature recommended for preventing acrylamide formation. For french fries, color was the main criteria when deciding the end-point of frying (85.3%). Although a golden color was preferred by respondents (87.3%), color guidelines are recommended in order to unify the definition of "golden." The results conclude that habits of the Spanish population are in line with recommendations to mitigate acrylamide during french fry preparation. Furthermore, these habits do not include practices that risk increasing acrylamide formation. Nevertheless, educational initiatives tailored towards consumers would reduce the formation of this contaminant and, consequently, exposure to it in a domestic setting.

Keywords: consumers; domestic habits; french fries; frying habits; households; oil; acrylamide

---

1. Introduction

Chemical process contaminants are substances formed when foods undergo chemical changes during processing, including heat treatment, fermentation, smoking, drying and refining [1]. Although necessary for making food edible and digestible, heat treatment can have undesired consequences leading to the formation of heat-induced contaminants such as acrylamide [2]. Acrylamide is a chemical process contaminant formed when foods containing free asparagine and reducing sugars are cooked at temperatures above 120 °C in low moisture conditions [3]. It is mainly formed in baked or fried carbohydrate-rich foods as the relevant raw materials contain its precursors. These include cereals, potatoes and coffee beans. In 1994, acrylamide was classified by the International Agency for Research on Cancer as being probably carcinogenic to humans (group 2A) [4]. In 2015, the European Food Safety Authority (EFSA) confirmed that the presence of acrylamide in foods is a public health concern, requiring continued efforts to reduce its exposure [3].

Fried potato products are the main contributor to total dietary acrylamide exposure, especially amongst young people [3]. The recent 2017/2158 Regulation [5] on benchmark levels for reducing the presence of acrylamide in foods and the acrylamide toolbox compiled by Food Drink Europe [6] include specific mitigation strategies for decreasing the presence of acrylamide in foodstuffs. The category of fried potato products is divided between two subcategories. The first category describes potato-based snacks, whilst the second describes french fries and other cut potato products. When fried potato products are made from fresh potatoes, mitigation measures are focused on the following factors: selection of suitable potato varieties, acceptance criteria based on quality, potato storage and transport, recipe and process design. In addition, information for end users can be supplied in packaging if potatoes are intended to be fried at home. These recommendations are aimed at controlling the precursor levels in fresh tubers (mainly reducing sugar content), frying temperature and the color of the final product. Further, they are fundamentally focused on the industrial sector [5,6]. As a result of following these measures, acrylamide levels in potato chips have decreased in recent years, demonstrating that mitigation strategies are being successfully applied to industrial potato crisps [7–9]. However, these strategies cannot be directly extended to private domestic settings as the main variables accounting for the cooking process vary between households and even between individuals [10]. Consumers may significantly influence their dietary acrylamide exposure through their purchase choices and the selection of culinary methods for cooking food, amongst other factors [11,12]. The EFSA has reported that home-cooking behaviors for potato frying lead to variations of up to 80% total dietary acrylamide exposure.

Several initiatives have been driven by different national food safety authorities with the aim of helping consumers understand how they can minimize acrylamide exposure when cooking at home. Examples of this include the "go for gold" campaign by the Food Standards Agency (FSA) [11] and the "golden but not toasted" campaign by the Spanish Food Safety Agency [13]. Both campaigns intended to educate audiences on how to identify the golden color that is characteristic of a healthier fry and lower acrylamide content in fried potatoes. Despite these campaigns, the population may not be applying mitigation measures during the preparation of french fries at home. They may even be unaware that acrylamide is an issue and that its formation can be reduced during frying. In June 2019, the EFSA published the last special Eurobarometer on risk perceptions, which included the results of a survey of 27,655 respondents from different social and demographic groups around Europe [14]. The survey showed that when considering commonly known food safety related topics, European citizens were most concerned about antibiotic residues; hormones or steroids in meat (44%); pesticide residues in food (39%); environmental contaminants in fish, meat or dairy products (37%) and additives such as colorants and preservatives or flavors used in food or beverages (36%). Process contaminants in foods, including acrylamide, were not listed amongst these topics.

Within this framework, the aim of the present work was to explore domestic practices for the preparation of french fries in Spanish households using a survey of domestic culinary habits and consumer preferences. The degree of compliance with mitigation strategies for reducing acrylamide formation in french fries was then evaluated, whilst also identifying unwanted practices that could be modified through re-education to favor healthier frying habits amongst the Spanish population. To our knowledge, this is the first survey conducted in a Spanish population with a focus on french fry preparation habits.

2. Materials and Methods

2.1. Study Design

The experimental design was based on similar studies in the literature [15–17]. The questionnaire was supervised by researchers from the Food Innovation research team at the Polytechnic University of Valencia (Valencia, Spain). Firstly, the survey was piloted with 20 individuals via one-on-one interviews. This was done to identify specific focal points for questionnaire development and to make the necessary adjustments based on respondent feedback. Following this, the questionnaire was drafted and validated with 35 subjects. The survey included thirty-two questions divided into seven different topics. These were:

i.   Socio-demographic characteristics: gender, age group, nationality, type of household, number of individuals at home and number of individuals under 18 years old at home.
ii.   Culinary habits: cooking experience.

iii. French fry consumption: characteristics of potatoes intended for frying (fresh or frozen, par-fried), frequency of french fry consumption and the way in which they are consumed.

iv. Fresh potato characteristics: place of purchase and type of fresh tubers (in-season or stored, washed or unwashed, bulk or bagged, labelled as "special for frying," etc.), geographical origin, botanical variety and place of storage at home.

v. Practices at the pre-frying stage: peeling, washing, soaking, adding salt and cutting preferences.

vi. Practices at the frying stage: kitchen utensil, frying oil characteristics, frying temperature, subjective ratio between the amount of food with respect to the dimensions of the frying utensil, defrosting prior to par-frying potatoes or frying from frozen and frying cycles.

vii. Post-frying practices: criterion for establishing the frying end-point, color and texture preferences; method for removing oil from fried food; criteria for stopping oil use in other frying cycles; cleaning procedures for oil reuse and oil storage.

Questions were structured according to check boxes with unique or multiple possible responses. The questionnaire included a brief introduction about the aim of the study and participants gave their consent to use their data in the research (Supplementary Material: Survey on household habits in relation to frying potatoes).

2.2. Data Collection

The present study was conducted with 730 respondents, using both online ($n = 421$) and paper-based ($n = 309$) questionnaires. A web-based survey was launched via an internet platform through Google form links (Melbourne, Australia), and paper-based questionnaires were directly distributed to respondents. Questionnaire data were included when more than 75% of the survey was completed and no more than two questions per topic were missing. Both datasets were merged for analysis since the approach through which information was compiled was not a limiting source of variability for analysis. The selection of participants was limited to adults who consume or prepare french fries at home. The main channels for questionnaire distribution were schools, consumer associations, universities, food associations, research centers, culinary centers, etc. Participants agreed to the anonymous use of their data for the study.

2.3. Data Processing

Outputs from the online and paper-based questionnaires were merged and compiled in an excel spreadsheet format. Data processing was performed using SPSS version 23.0 (SPSS Inc., Chicago, IL, USA). Categorical data were expressed as frequencies and percentages.

3. Results and Discussion

3.1. Characteristics of the Participants

The present study aimed to evaluate the adhesion of potato frying habits in Spanish households to recommendations for mitigating acrylamide formation provided by the European 2017/2158 Regulation [5]. It has to be pointed out that the purpose was not to give advice on acrylamide and frying practices or to assess the population's knowledge of these issues but to evaluate the domestic habits for the preparation of french fries and to compare them with the recommendations. Figure 1 depicts the critical action points for reducing acrylamide formation during the preparation of fried potatoes from fresh tubers according to three stages (pre-frying, frying, post-frying). An ad-hoc survey of culinary practices for the preparation of fried potatoes in households was conducted with Spanish adults. To obtain a representative and diverse number of respondents and minimize response bias, the survey included both online and paper-based questionnaires [18]. The web-based format allows for a broader dissemination, being less time demanding and less expensive, whilst paper-based questionnaires can be distributed amongst people who are not familiar with the internet or have limited access to digital platforms.

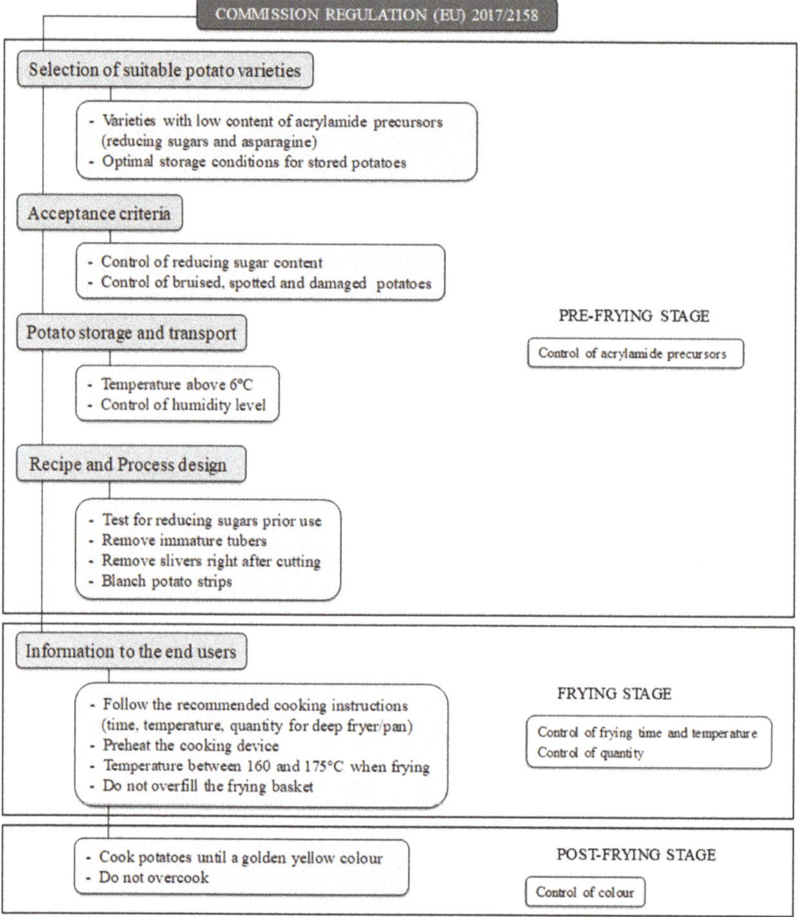

Figure 1. Mitigation measures referred to french fries made from raw potatoes according to the 2017/2158 Commission Regulation (EU) adapted for domestic habits.

For the cross-sectional population survey, a cohort of seven hundred and forty-eight volunteers was recruited from 47 of the 50 provinces of Spain. Eighteen participants did not complete the paper-based questionnaire and their data were withdrawn. Thus, the final number of responders included in the study was 730. Four hundred and twenty-one individuals filled out the online questionnaire and 309 completed the paper version. Participants were adults who usually prepare or consume french fries at home. The sociodemographic characteristics of the participants are summarized in Table 1. Despite that currently, 11% of the population living in Spain is foreign [19], 97.8% of the surveyed were of Spanish nationality. Participants were mostly female (73.2%), and aged between 36 and 55 years old (54.1%). With regards to the level of culinary experience, 72.3% of women considered themselves to have high expertise in contrast to 52.1% of men. This distribution corroborates gender differences reported in the ENHALI-2012 study [20], identifying a higher participation of women in the domestic activity of food preparation. Households were constituted by two (29.7%), three (26.0%) or four or more individuals (34.3%), with most not being comprised of any individual aged under 18 years

(59.5%). The majority of participants live with a partner (70.7%), with 45.1% living with children and relatives, and 25.6% being couples without children. The household profile is consistent with the typology described for the Spanish population, with smaller and one-person households made up of a young person or an independent adult predominating [21].

Table 1. Sociodemographic characteristics of participants (n = 730).

Characteristic	n	(%)
Gender		
Male	190	(26.0)
Female	534	(73.2)
Missing	6	(0.8)
Age group		
18–35	209	(28.6)
36–55	395	(54.1)
56–65	86	(11.8)
>65	38	(5.2)
Missing	2	(0.3)
Nationality		
Spanish	714	(97.8)
Other than Spanish	8	(1.1)
Missing	8	(1.1)
Individuals < 18 at home		
Yes	290	(39.7)
No	434	(59.5)
Missing	6	(0.8)
Individuals/home		
1	63	(8.6)
2	217	(29.7)
3	190	(26.0)
4	210	(28.8)
5	33	(4.5)
>5	7	(1.0)
Missing	10	(1.4)
Type of household		
Single	61	(8.4)
Single with children	30	(4.1)
Shared apartment	27	(3.7)
Couple without children	187	(25.6)
Couple with children	313	(42.9)
Couple, children and relatives	16	(2.2)
With parents	26	(3.6)
With parents and siblings	47	(6.4)
Other	12	(1.6)
Missing	11	(1.5)

Number of cases (n).

3.2. French Fry Consumption

Most respondents (85.3%) usually buy fresh tubers to prepare french fries at home. In total, 21 (2.9%) reported preparing only frozen par-fried potatoes, with 84 (11.5%) using both foods indistinctly (Table 2). This finding agrees with the data described in the food consumption report in Spain 2018. This indicated that fresh tubers and frozen par-fried potatoes represent 72.7% and 3.4%, respectively, of all potatoes bought by the Spanish population [21]. A total of 43.5% of respondents affirmed consuming french fries several times a month (monthly), whilst 35.9% reported a consumption of several times a week (weekly). A lower percentage eats these products only on exceptional occasions (19.7%) and only two individuals indicated a daily consumption (0.3%). With regards to consumption preferences, french fries make up the side dish for other foods (97.1%) such as meat (79.7%); fish (23.8%); vegetables (8.9%) and other foods including fried eggs, sausages, croquettes, etc. (47.6%).

Potato chips or french fries are a cause of concern due to their high fat and energy content. This is associated with a higher incidence of diseases such as obesity, high blood pressure and/or hypercholesterolemia [22]. For this reason, although frying is an ancestral and popular culinary technique and a typical way of cooking in the Mediterranean diet [23], nutritional recommendations limit this culinary practice to prevent its associated health complications. The relationship was evaluated between french fry consumption and both age of and household type. The most significant difference was observed in people older than 65 years of age who usually consume potatoes several times a week (weekly), with the frequency of consumption in other age groups predominantly being several times a month (monthly) (Figure 2A). Households with a higher potato consumption were typically formed by couples with children, followed by couples without children. These groups predominantly reported a monthly or weekly consumption (Figure 2B). Similarly, in Spain, a higher potato consumption corresponds to households made up of middle-aged and older children, single-parent households and households made up of adult couples without children [21].

3.3. Fresh Potato Characteristics

The survey showed that fresh tubers are mainly bought in supermarkets (48.1%) and neighborhood grocery stores (44.1%), followed by hypermarkets (29.3%) and markets (16.2%) (Table 2). This agrees with purchase habits described for the global Spanish population [21]. Respondents are generally interested in knowing the geographical origin and/or botanical variety of the potato, although 38.1% are not interested in this aspect. Some of the common varieties used in Spain to prepare french fries are Monalisa, Caesar, Milva and Agria [12]. Potatoes are normally bought in bulk rather than bagged (36.1 vs. 28.1), and are washed (43.3%) and fresh harvested (in-season) (49.1%). However, some respondents expressed not being concerned about the type of potato purchased in relation to its presentation (6.0%) and seasonality (24.1%).

Five hundred and seventeen individuals (70.8%) indicated that they do not buy "special frying potatoes," with a further 98 (13.4%) not being aware of this type of classification. According to commercial quality norms for potatoes for consumption in the Spanish market, commercial potatoes must be identified according to product type (variety, uncalibrated/calibrated, in-season/stored), origin (country and, optionally, regional or local production area or national denomination) and commercial characteristics (category, net weight, batch, caliber/size, recommended culinary use, etc.). For bagged potatoes, the name, trade name or designation of the packer and/or shipper or seller should also be identified [24]. Regarding the place of storage, 79.6% store potatoes indoors, whilst 4.0% keep their potatoes both indoors and outdoors (Figure 3A) The UK Food Standards Agency (FSA) drew up a report which aimed to provide information on actual domestic cooking and french fry preparation practices in the UK. The report indicated that the majority of individuals in the UK usually store their potatoes outside of the fridge, such as in a kitchen cupboard, garage or utility room, prior to preparing and cooking them [11]. This is similar to the Spanish habits observed in the present study.

Table 2. French fry consumption and characteristics related to the use of fresh potatoes to prepare french fries in a Spanish population (number of cases, (%)).

	French Fry Consumption								
	fresh		frozen		not distinguished				missing [1]
Type	623	(85.3)	21	(2.9)	84	(11.5)			2	(0.3)
	daily		weekly		monthly		exceptionally		missing [1]
Frequency of consumption	2	(0.3)	262	(35.9)	318	(43.5)	144	(19.7)	4	(0.6)
	only french fries		as an accompaniment						missing [1]
Consumption	19	(2.6)	709	(97.1)					2	(0.3)
	Fresh Potatoes								
	neighborhood stores		local markets		supermarkets		hypermarkets		missing [1]
Place of purchase [2]	322	(44.1)	118	(16.2)	351	(48.1)	214	(29.3)	0	(0.0)
	only origin		only variety		both		no		missing [1]
Origin/variety	103	(14.1)	84	(11.5)	264	(36.2)	278	(38.1)	1	(0.1)
	yes		no				unknown [3]		missing [1]
Special for frying	113	(15.5)	517	(70.8)			98	(13.4)	2	(0.3)
	bulk		bagged		both		unnoticed [4]		missing [1]
Packaging	268	(36.7)	205	(28.1)	255	(34.9)	0	(0.0)	2	(0.3)
	unwashed		washed		both		unnoticed [4]		missing [1]
Presentation	99	(27.3)	316	(43.3)	171	(23.4)	44	(6.0)	0	(0.0)
	in-season		stored		both		unnoticed [4]		missing [1]
Seasonality	358	(49.1)	31	(4.2)	164	(22.5)	176	(24.1)	1	(0.1)

[1] Not answered; [2] multiple possible answers; [3] volunteer does not care about this; [4] volunteer cannot be precise.

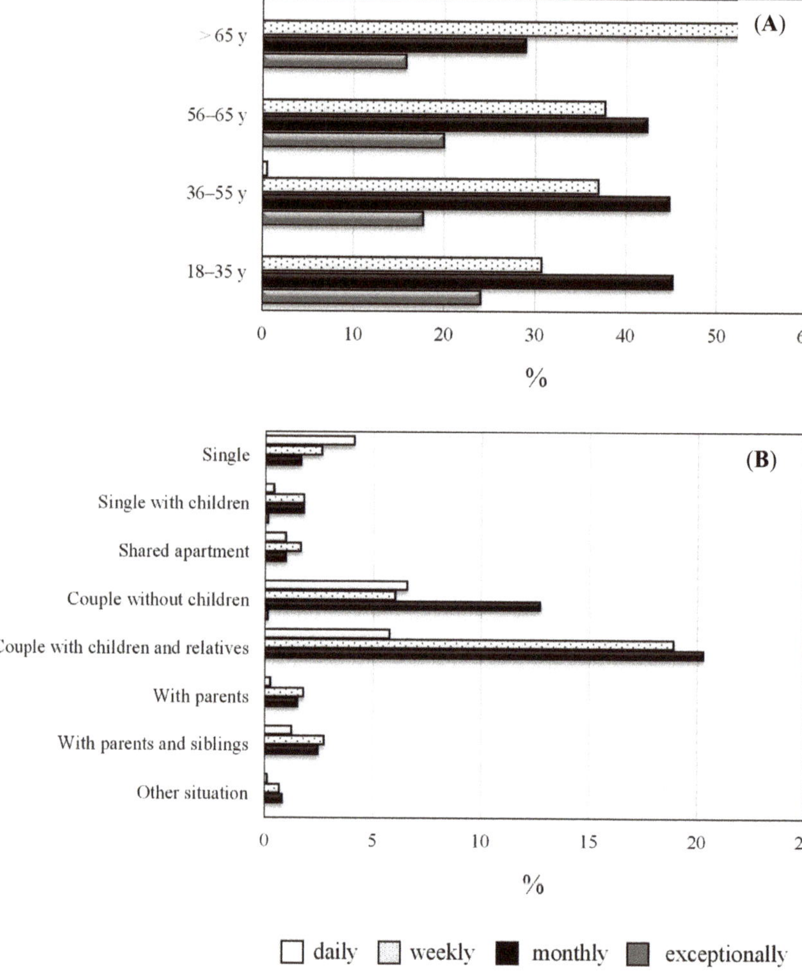

Figure 2. Distribution of french fry consumption according to participants' age (expressed in relative amounts within each age group) (**A**) and the type of household (expressed in absolute amounts for all respondents (**B**): □ daily, ▨ weekly, ■ monthly and ▪ exceptionally).

For the preparation of fried potato products based on raw potatoes, the 2017/2158 Regulation recommends the selection of suitable potato varieties. This urges a reducing sugar (fructose and glucose) and asparagine content that is as low as possible for regional conditions [5] (Figure 1). Asparagine and reducing sugars are well-known acrylamide precursors, with reducing sugars being the limiting factor of acrylamide-forming potential in potato products [25,26]. Controlling reducing sugar levels is, therefore, currently the primary measure employed by the industry to reduce acrylamide concentration in french fries. This is achieved by selecting potato varieties with a low content of reducing sugars, ensuring that tubers are mature at the time of harvesting, controlling storage conditions and managing humidity to minimize senescent sweetening [6]. Due to the seasonality of this commodity, reducing sugar content varies depending on whether the potato is fresh harvested or stored [27]. It is well documented that temperatures below 6 °C for long-term storage, without an adequate re-conditioning step before frying, lead to senescent sweetening as a result of starch hydrolysis [28]. The higher reducing sugar content of stored potatoes promotes the formation of higher levels of acrylamide during the frying process [6]. However, all of these practices must be controlled at both farming and industrial stages in order to provide consumers with optimum potato tubers for helping to mitigate acrylamide formation in the domestic environment. The most important consumer's decision when purchasing potatoes for the preparation of french fries is to select those that are fresh harvested, whenever possible, and store them indoors, since temperatures inside the home will not promote the mentioned sweetening. On the other hand, although potatoes labelled as "special for frying" could be expected to be the most suitable potatoes for frying, previous research has demonstrated that this commercial label is not always adequate when guiding consumers in the preparation of french fries with low acrylamide content [12].

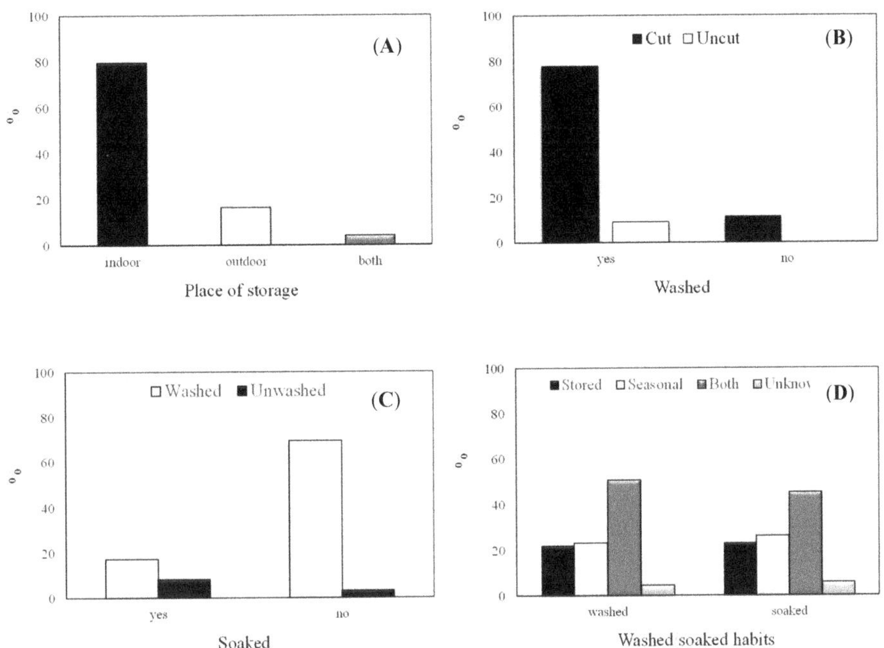

Figure 3. Handling habits of fresh tubers: place of storage (**A**), washing-related habits (cut or uncut potatoes) (**B**), soaking-related habits (washed or unwashed potatoes) (**C**), washing and soaking-related habits (stored and seasonal potatoes) (**D**).

It might be concluded that the Spanish population tends to adhere to acrylamide mitigation strategies relating to storage conditions (Figure 4). However, more information should be provided for consumers to improve their selection of raw materials. This is especially the case for those who do not take the type of potato purchased on the market into consideration, as the acquisition of stored potatoes probably promotes higher acrylamide formation and, thus, a higher reducing sugar content. In addition, use of the "special for frying" label should be supervised by the relevant food authority in order to ensure products exhibit low levels of precursors, in addition to the suitable raw materials intended for frying.

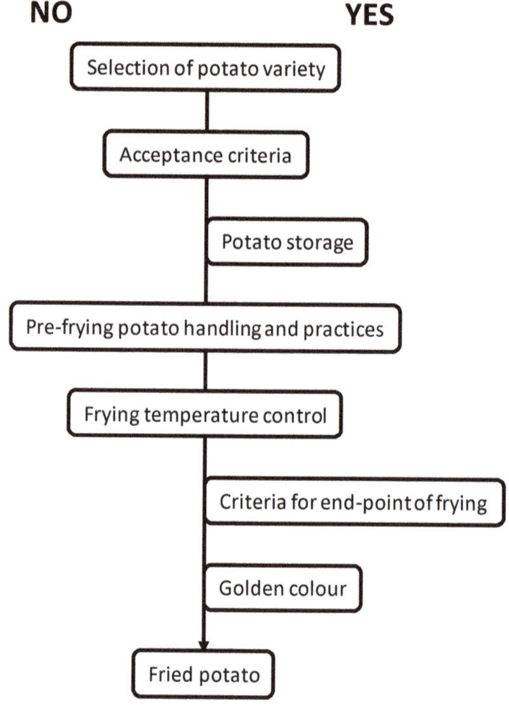

Figure 4. Adherence of the Spanish population to EU recommendations for the mitigation of acrylamide formation in potatoes fried from raw potatoes in a domestic setting [5].

3.4. Practices Related to the Pre-Frying Stage

Before frying, most participants peel the potatoes (97.5%) (Table 3) and wash them after peeling (86.8%). Only 43 respondents (5.9%) usually wash tubers before peeling. Amongst the volunteers who wash potatoes ($n = 634$), 66 cut them after washing, whilst the remaining participants do so beforehand (Figure 3B). Most volunteers ($n = 531$) do not soak the potatoes; however, 508 do wash the potato tubers after peeling. In contrast, 126 individuals both wash and soak the potatoes (Figure 3C), with less than 15 min being the most commonly applied soaking time (126 respondents) (Table 3). Both washing and soaking habits are indistinctly applied with stored and in-season tubers (Figure 3D). In order to decrease the sugar content of fresh tubers, especially in the case of stored potatoes, practices such as washing, soaking and/or blanching should be applied [29,30]. The European Commission recommends washing and soaking for 30 min to 2 h in cold water, soaking for a few minutes in warm water or blanching [5] (Figure 1). Although respondents mostly wash potatoes before frying, soaking practices should be included as common habits during french fry preparation in Spain. These practices should be especially promoted when using stored potatoes, as they lead to improved compliance with pre-frying potato handling and practice recommendations (Figure 4). It should be noted that consumers may wash potatoes for reasons other than to mitigate acrylamide formation. Reasons may include heritage habits, hygiene purposes or simply to prevent browning whilst other elements of the meal are prepared. This observation was also made in the FSA report [11].

Table 3. Habits for frying potatoes in a Spanish population (number of cases, (%)).

	Pre-Frying Stage									
	yes		no						missing [1]	
Peeled	712	(97.5)	2	(0.3)					16	(2.2)
	yes		no						missing [1]	
Washed	634	(86.8)	82	(11.3)					14	(1.9)
	yes		no						missing [1]	
Soaked	185	(25.4)	531	(72.7)					14	(1.9)
	<15 min		15–30 min		>30 min					
If soaking, duration	126	(17.3)	43	(5.9)	16	(2.2)				
	before frying		during frying		after frying		no		missing [1]	
Salt	263	(36.1)	39	(5.3)	347	(47.5)	79	(10.8)	2	(0.3)
	strips		cubes		chips		irregular		slices	
Type of cut [2]	642	(87.9)	142	(19.5)	57	(7.8)	85	(11.6)	158	(21.6)
	Frying Stage									
	frying pan		electric fryer		both		other		missing [1]	
Kitchen appliance	577	(79.0)	67	(9.2)	63	(8.6)	23	(3.2)	0	(0.0)
	olive		sunflower		both		other		missing [1]	
Type of oil	574	(78.7)	128	(17.5)	20	(2.7)	7	(1.0)	1	(0.1)
	taste		price		health		performance		appliance	
Criteria for oil selection	375	(51.3)	116	(15.9)	468	(64.1)	132	(18.1)	25	(3.4)
	yes		no		unknown [3]				missing [1]	
Special for frying	51	(7.0)	535	(73.3)	144	(19.7)			0	(0.0)
	<half		half		>half		full		missing [1]	
Potato/appliance surface	59	(8.1)	187	(25.6)	247	(33.8)	235	(32.2)	2	(0.3)

Table 3. *Cont.*

	yes		no		not consumed				missing [1]	
Defrosts frozen potatoes [4]	6	(0.8)	100	(13.7)	621	(85.1)			3	(0.4)
	one		two						missing [1]	
Frying cycles	696	(95.3)	22	(3.0)					12	(1.7)
Post-Frying Stage										
	soft		crunchy-soft		crunchy				missing [1]	
Texture	103	(14.1)	603	(82.6)	24	(3.3)			0	(0.0)
	paper		rack		both		none		missing [1]	
Removal of oil	424	(58.1)	114	(15.6)	154	(21.1)	36	(4.9)	2	(0.3)
	strainer		paper		decantation		no cleaning		missing [1]	
Cleaning	439	(60.1)	39	(5.3)	77	(10.6)	164	(22.5)	11	(1.5)
	used appliance		closed container		open container		no reusing		missing [1]	
Storage	133	(18.2)	384	(52.6)	102	(14.0)	103	(14.1)	8	(1.1)

[1] not answered; [2] multiple possible answers; [3] volunteer does not care about this; [4] only for frozen par-fried potatoes.

Other aspects to mention in relation to the pre-frying stage include the addition of salt and the type of potato cut. Salt is mainly added after frying (47.5%) and potatoes are mainly cut into strips (87.9%), followed by slices (21.6%) and cubes (19.5%). In response to this question, participants could provide multiple answers. For this reason, the sum of partial percentages is higher than 100%. Although neither the relevant regulation nor the acrylamide toolbox mention any indication about adding salt before frying, the addition of calcium salts is recommended for the preparation of dough-based potatoes in order to reduce the pH level [5,6]. The presence of NaCl in the food matrix has been described as decreasing the acrylamide content following heat treatment [31]. Thus, as long as the amount of added salt is not high and adheres to health recommendations for salt consumption [32], the population should be encouraged to add salt before frying. With regards to cutting preferences, given that acrylamide is formed on the surface of fried food, controlling thickness is an important factor for preventing its formation. For french fries, it is better to cut in strips than in slices and, in addition, creating thicker strips of potato may reduce acrylamide in french fries [6]. In this sense, several authors have observed an inverse trend between acrylamide levels and french fry thickness [33,34]. The FSA study [11] highlighted, in an English sample, that homemade potato items tended not to be finely chopped, avoiding the risk of additional acrylamide exposure through greater surface area to volume ratios. Similarly, Spanish habits with respect to the type of cut selected for french fries appear to be appropriate, although thickness should be more thoroughly examined.

3.5. Practices Related to the Frying Stage

Practically all respondents (95.3%) carry out a single frying cycle (Table 3). More than three quarters generally use a frying pan to fry potatoes (79.0%), with 9.2% remarking that the electric fryer was the preferred utensil and 8.6% indicating that they used both types of cookware indistinctly (Table 3). These results contrast those reported by Romero et al. [15], who reported that Spanish University students mainly used electric fryers to prepare french fries (71.3%). Olive oil is the most widely used oil for frying potatoes (indicated by 78.7% of volunteers), with a much smaller percentage selecting sunflower oil as the preferred oil (17.5%). This reflects other profiles reported in previously conducted studies in Spain [35,36]. Respondents' oil choices are conditioned by their related health properties (468 subjects), taste (375 subjects), oil stability (132 subjects), price (116 subjects) and, to a lesser extent, the appliance used for frying (20 subjects). The majority of participants (73.3%) do not consume oil labelled as "special for frying," with 144 even being unaware of the existence of these oils. The use of olive oil for frying is associated with healthy habits as it improves the saturated/monounsaturated/polyunsaturated fatty acid profile of food and enriches its fat-soluble vitamin and antioxidant compound content [37]. However, with regards to acrylamide formation, no significant differences have been observed following the preparation of french fries in different edible oils [33]. Recommendations of the acrylamide toolbox and relevant regulation do not make any type of reference to the type of oil to be used for frying, though they do specify the maximum temperatures advised for the oil (Figure 1). Temperature is one of the main factors to determine acrylamide formation in processed foods and, consequently, its control is absolutely essential [33]. Frying temperatures are recommended to be below 175 °C, and as low as possible at all times whilst considering food safety requirements. In addition, it is recommended to preheat the cooking utensil in order to reduce frying time [5,6]. All volunteers preheat the oil and indicated different criteria for checking that oil temperature is ready for frying: adding a potato and observing its behavior (47.5%), smoke emission (smoke point) (18.9%) and controlling the thermostat of the electric fryer (7.9%). At the same time, 24.5% reported not having precise control, adding the food to the frying pan after the oil has warmed up for a period of time (without control) (Figure 5A). The majority of Spanish participants (80.8%) stated that they control the frying temperature by adjusting the thermostat (in the case of using an electric fryer) or the power of the gas flame, glass ceramic or induction plate (in the case of a frying pan) (Figure 5B). From the remaining sample, 68 people always select the maximum temperature, whilst 59 do not control this aspect. Although the majority of individuals reported exercising control of the frying temperature, a temperature below 175 °C cannot be ensured (Figure 4). Thus, recommendations on the optimal frying temperature for potatoes in order to prevent acrylamide formation should be addressed with the population. Frying temperature may be substantially lower owing to the strong cooling that results from heating the potatoes and water evaporation, further depending on the amount of potato added to the volume of frying oil [38]. Information about the food/frying oil ratio was not obtained in the present survey; however, respondents made inferences based on the surface area of the appliance used. In this sense, the amount of food intended for frying typically amounted to more than half of the dimension of the frying surface (frying pan or frying basket) (33.8%) (Table 3). Thus, not only should recommendations be made for temperature control, but potato/oil ratio should be also considered.

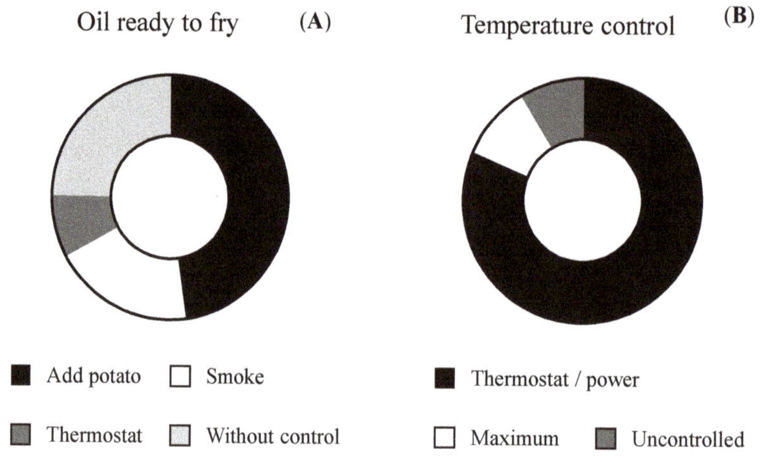

Figure 5. Selection of frying starting point (**A**) and temperature control during french fry preparation (**B**).

Only 14.5% of participants (*n* = 106) reported consuming frozen par-fried potatoes. Of these, 100 subjects do not defrost the product before frying, following the indications provided on the label of these products. This result could be considered a good practice as the relevant regulation suggests that cooking instructions should be followed for frozen potato products [5].

3.6. Practices Related to the Post-Frying Stage

The literature describes a correlation between color and acrylamide formation in french fries [12,39–41]. Thus, the decision of food handlers regarding the final color will influence acrylamide exposure from french fries. It is recommendable to cook until a golden (yellow) color is achieved and to avoid overcooking [5,6] (Figure 1). Several educational initiatives have been undertaken to orientate consumers so that they associate the extent of browning in french fries with mitigation strategies for acrylamide at home. Examples include the "go for gold" campaign launched by the UK Food Standards Agency [11] and the slogan "golden but not toasted" by the Spanish Food Safety Agency [13]. Participants in the present study showed an appropriate compliance with recommendations when selecting the color of french fries as the main criteria for choosing the frying end-point (85.3%), with only a minority of respondents stopping the procedure by tasting a sample (11.9%) or when the fried potato stops bubbling in the oil (1.2%) (Figure 6A). Despite golden being the preferred color (87.3%) (Figure 6B), the distribution of color guides providing guidance on the optimal combination of color and low acrylamide levels should not be discarded since previous research by our team has identified the necessity of a clear definition of "golden" amongst consumers [41]. In agreement with these observations, visual checking of color and appearance was the dominant way to decide whether a product was "ready" as desired within a sample of English consumers [11].

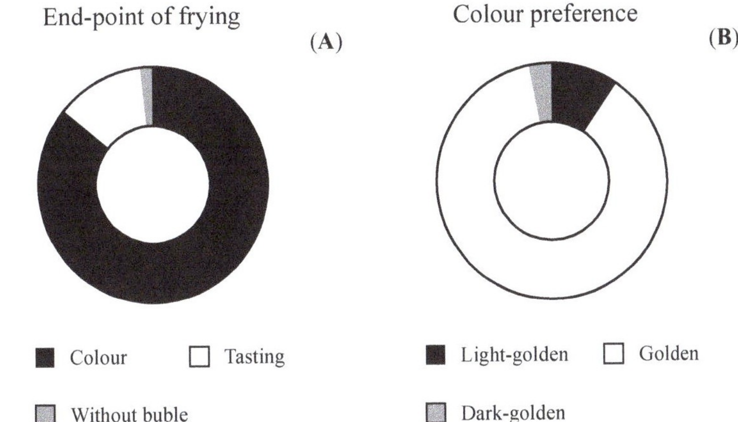

Figure 6. Criteria to decide the frying end-point (**A**) and color preferences in relation to french fries (**B**).

Another aspect collected in the survey refers to the texture of french fries, with crunchy on the outside and soft on the inside being the preferred characteristic (82.6%) (Table 3). Further, approaches to removing oil from the fried product were considered, with this mainly performed through the application of absorbent paper (58.1%) or paper and rack (21.1%). With regards to the use of frying oil, it is known that oils deteriorate during frying. This leads to changes in the fatty acid composition and retention of other oil-degradation products in fried foods [42]. This fact becomes even more relevant with repeated use. It is, therefore, important to control the number of times that frying oil is reused. In the present survey, amongst frying pan users, 25.8% utilize fresh oil whereas 63.1% reuse oil 2–4 times. A minority of individuals stated reusing oil 4–8 times (8.8%) or more than 8 times (2.3%) (Figure 7). The amount of oil used in the frying pan tended to be small, with alterations appearing quickly. Oil should therefore be changed more frequently. In contrast, electric fryers use a greater volume of oil, leading to more diluted alteration products and, therefore, allowing more foodstuffs to be fried [37]. Only 130 respondents of the present survey used this appliance; four of these indicated that they always use fresh oil, with a higher proportion reusing the oil: 4–8 times (41.5%), more than 8 times (37.7%) and 2–4 times (17.7%).

Factors considered by participants to determine changes to the frying oil are shown in Figure 7B. As shown, the main factors are organoleptic in nature, such as the darkening of the oil (32.0%) and presence of sediments (20.1%). Taste (3.8%) and viscosity (4.2%) were reported to a lower extent. In contrast, 27.8% consider the number of frying cycles used to be the main criterion when determining an oil change. In this sense, Romero et al. [15] reported that Spanish University students usually changed the oil in the electric fryer after a number of frying processes: <5 times (9%), 5–10 times (21%), 11–20 times (25%) and >20 times (8%). Similarly, Gatti et al. [16] indicated that Argentinian adults tended to reuse oil used for domestic practices once (28%), twice (40%), three times (4%) or until some extent of alteration was observed (28%), although they did not specify the type of appliance used. Spanish volunteers mainly clean the oil by filtering it with a strainer (60.1%) and storing it in a closed container (52.6%), open container (14.0%) or in the same frying container (13.2%) (Table 3). Similar habits have been described for adults from Argentina. In this case, the main technique was to use paper to remove excess oil from the food (84%); filter the oil before storage (52%) and store it in the same container (60%), in a glass container (20%) or in a plastic container (20%) [16]. Oil cleaning is also mentioned in acrylamide mitigation strategies during potato frying since removing fines and crumbs is recommended for maintaining frying oil quality [5]. Since frequent use of the same oil for frying potatoes was reported as a household habit of respondents, more emphasis on the importance of oil renewal and its cleaning should be made in communications within the Spanish population.

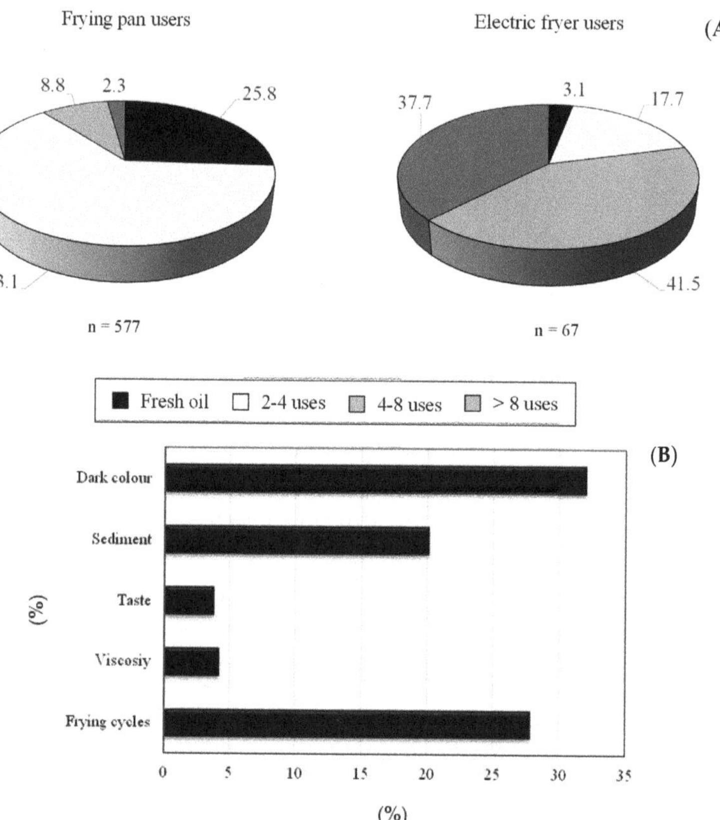

Figure 7. Frequency of use of frying oil (**A**) and reason for changing (**B**).

4. Conclusions

In the present study, a survey of domestic practices for cooking and preparing french fries was conducted within the Spanish population. The aim was not to give advice on acrylamide or frying practices but to provide information on domestic habits in relation to french fry preparation, identify which consumer practices may influence acrylamide formation and evaluate compliance with the acrylamide mitigation strategies described by the European Regulation. Results indicated that the Spanish population mainly prepares french fries from fresh tubers acquired in supermarkets (48.1%), preferably in-season potatoes (49.1%), purchased in bulk (36.1%), washed (43.3%) and stored indoors (79.6%). These habits, together with frequent washing of potatoes before frying, are adequate for controlling acrylamide precursor levels (mainly reducing sugar content). However, consumers should be more mindful when selecting the raw material. This is especially the case for those who do not consider the type of potato purchased at market. Improved practices would also include more frequently soaking potatoes to reduce the sugar content of the fresh tubers, particularly during periods when stored potatoes are the only ones available on the market.

Following the typical Mediterranean culinary practices, olive oil was the most commonly used frying oil, followed by sunflower oil, whilst frying pans were the most common utensils used for the frying process, leading individuals to more frequently replace cooking oil. Although most respondents reported controlling frying temperature, advice relating to the maximum temperature recommended for preventing acrylamide formation in french fries should be addressed. On the other hand, examining the color of french fries as the main criteria for deciding the frying end-point was often reported, which coincides with one of the main tips for controlling acrylamide levels and, consequently, exposure to it. In this regard, although a golden color was preferred by respondents, color guidelines should be distributed in order to clarify the definition of "golden." The population should also be instructed about the introduction of practices in relation to adding salt prior to frying, the type of cut, amount of potatoes in the appliance and cleaning oil after frying in order to understand all possible strategies for acrylamide mitigation at home.

From these results, it might be deduced that culinary domestic habits of the Spanish population are in line with recommendations for mitigating acrylamide during the preparation of french fries. Furthermore, they do not include any practices that increase the risk of acrylamide exposure. Nevertheless, some actions could be mentioned in order to control its presence in this food: (1) educational initiatives including information provision targeting improved domestic behaviors for the promotion of healthy frying habits should be promoted; (2) information should be specially focused on the importance of the selection of suitable fresh tubers; (3) on the washing and soaking of potatoes before frying; and (4) on the control of frying temperatures below 175 °C, avoiding excessive browning of food. All these measures will effectively prevent acrylamide formation and, consequently, will reduce its exposure.

Supplementary Materials: The following are available online at http://www.mdpi.com/2304-8158/9/6/799/s1, Supplementary material: Survey on household habits in relation to frying potatoes.

Author Contributions: Conceptualization, M.M., C.D.-A. and F.J.M.; methodology, M.M.; software, M.M.; validation, M.M., C.D.-A. and F.J.M.; formal analysis, M.M.; investigation, M.M., C.D.-A. and F.J.M.; resources, F.J.M.; data curation, M.M., C.D.-A. and F.J.M.; writing—original draft preparation, M.M.; writing—review and editing, M.M., C.D.-A. and F.J.M.; supervision, M.M., C.D.-A. and F.J.M.; funding acquisition, M.M., C.D.-A. and F.J.M. All authors have read and agreed to the published version of the manuscript.

Funding: This research was partly funded by the Ministry of Economy and Competitiveness (Spain) under the projects SAFEFRYING (AGL2015-46234-R; MINECO) and ACRINTAKE (RTI2018-094402-B-I00, MCIU/AEI/FEDER, UE).

Acknowledgments: The authors thank J. Martinez-Monzó and P. García for their work in validating the questionnaire.

Conflicts of Interest: The authors declare no conflict of interest.

References

1. Bonwich, G.; Birch, C.S. European regulation of process contaminants in food. In *Mitigation Contamination from Food Processing*; Birch, C.S., Bonwick, G.A., Eds.; RSC: Cambridge, UK, 2019; pp. 1–16.

2. Van Boekel, M.; Fogliano, V.; Pellegrini, N.; Stanton, C.; Scholz, G.; Lalljie, S.; Somoza, V.; Knorr, D.; Jasti, P.R.; Eisenbrand, G. A review on the beneficial aspects of food processing. *Mol. Nutr. Food Res.* **2010**, *54*, 1215–1247. [CrossRef]

3. EFSA (European Food Safety Authority). Scientific opinion on acrylamide in food. *EFSA J.* **2015**, *13*, 4104.

4. IARC (International Agency for Research on Cancer). Some industrial chemicals. In *IARC Monographs on the Evaluation for Carcinogenic Risk of Chemicals to Humans*; IARC: Lyon, France, 1994; Volume 60, pp. 435–453.

5. EC (European Commission). Commission Regulation (EU) 2017/2158 of 20 November 2017 establishing mitigation measures and benchmark levels for the reduction of the presence of acrylamide in food. *OJEU* **2017**, *304*, 24–44.

6. FDE (FoodDrinkEurope). 2019. Available online: https://www.fooddrinkeurope.eu/uploads/publications_documents/FoodDrinkEurope_Acrylamide_Toolbox_2019.pdf (accessed on 12 April 2020).

7. Claeys, W.; De Meulenaer, B.; Huyghebaert, A.; Scippo, M.L.; Hoet, P.; Matthys, C. Reassessment of the acrylamide risk: Belgium as a case-study. *Food Control* **2016**, *59*, 628–635. [CrossRef]

8. Powers, S.J.; Mottram, D.S.; Curtis, A.; Halford, N.G. Acrylamide levels in potato crisps in Europe from 2002 to 2016. *Food Addit. Contam. Part A* **2017**, *34*, 2085–2100. [CrossRef]

9. Mesias, M.; Nouali, A.; Delgado-Andrade, C.; Morales, F.J. How far is the Spanish snack sector from meeting the acrylamide regulation 2017/2158? *Foods* **2020**, *9*, 247. [CrossRef]

10. Langiano, E.; Ferrara, M.; Lanni, L.; Viscardi, V.; Abbatecola, A.M.; De Vito, E. Food safety at home: Knowledge and practices of consumers. *Z. Gesundh.* **2012**, *20*, 47–57. [CrossRef]

11. FSA (Food Standards Agency). Acrylamide in the home: Home-Cooking Practices and Acrylamide Formation. 2017. Available online: https://www.food.gov.uk/research/research-projects/acrylamide-in-the-home-the-effects-of-home-cooking-on-acrylamide-generation (accessed on 14 April 2020).

12. Mesías, M.; Holgado, F.; Márquez-Ruiz, G.; Morales, F.J. Impact of the characteristics of fresh potatoes available in-retail on exposure to acrylamide: Case study for French fries. *Food Control* **2017**, *73*, 1407–1414. [CrossRef]

13. AECOSAN (Spanish Agency for Consumer Affairs, Food Safety and Nutrition). 2015. Available online: http://www.aecosan.msssi.gob.es/AECOSAN/web/seguridad_alimentaria/subdetalle/acrilamida.htm (accessed on 14 April 2020).

14. EFSA (European Food Safety Authority). Special Eurobarometer Wave EB91.3. 2019. Available online: http://www.efsa.europa.eu/sites/default/files/corporate_publications/files/Eurobarometer2019_Food-safety-in-the-EU_Full-report.pdf (accessed on 12 April 2020).

15. Romero, A.; Cuesta, C.; Sánchez-Muniz, F.J. Utilización de freidora doméstica entre universitarios madrileños. Aceptación de alimentos congelados fritos en aceite de oliva virgen extra, girasol y girasol alto oleico. *Grasas Aceites* **2001**, *52*, 38–44.

16. Gatti, M.B.; Cabreriso, M.S.; Chain, P.; Coniglio, A.; Manin, M.; Ciappini, M.C. Evaluación de la frecuencia de consumo de alimentos fritos y de las técnicas de frituras domésticas en adultos rosarinos. *Invenio* **2015**, *19*, 155–165.

17. Wood, K.; Carragher, J.; Davis, R. Australian consumer's insights into potatoes-nutritional knowledge, perceptions and beliefs. *Appetite* **2017**, *114*, 169–174. [CrossRef] [PubMed]

18. Van den Berg, M.H.; Overbeek, A.; van der Pal, H.J.; Versluys, B.; Bresters, D.; van Leeuwen, F.E.; Lambalk, C.B.; Kaspers, G.J.L.; van Dulmen-den Broeder, E. Using web-based and paper-based questionnaires for collecting data on fertility issues among female childhood cancer survivors: Differences in response characteristics. *J. Med. Internet Res.* **2011**, *13*, e76. [CrossRef]

19. Instituto Nacional de Estadistica. Cifras de Población Residente en España. Datos Provisionales a 1 de Enero de 2020. Available online: https://www.ine.es/dyngs/INEbase/es/operacion.htm?c=Estadistica_C&cid=1254736176951&menu=ultiDatos&idp=1254735572981 (accessed on 10 June 2020).

20. Díaz-Méndez, C.; García-Espejo, I.; Gutiérrez-Palacios, R.; Novo-Vázquez, A. *Hábitos alimentarios de los españoles*; Ministerio de Agricultura, Alimentación y Medio Ambiente: Madrid, Spain, 2013. Available online: http://www.besana.es/sites/default/files/estudio_habitos_alimentarios_de_los_espanoles.pdf (accessed on 10 April 2020).

21. MAPA (Ministry of Agriculture, Fisheries and Foods). Datos de Consumo Alimentario. Informe del Consume Alimentario en España. 2018. Available online: https://www.mapa.gob.es/images/es/20190807_informedeconsumo2018pdf_tcm30-512256.pdf (accessed on 10 April 2020).

22. Mellema, M. Mechanism and reduction of fat uptake in deep-fatfried foods. *Trends Food Sci. Technol.* **2003**, *14*, 364. [CrossRef]

23. Bastida, S.; Sánchez-Muniz, F.J. Frying: A cultural way of cooking in the Mediterranean diet. In *The Mediterranean Diet. An Evidence-Based Approach*; Preedy, V.R., Watson, R.S., Eds.; Academic Press: London, UK, 2015; pp. 217–236.

24. Boletin Oficial del Estado (BOE). Real Decreto 31/2009, de 16 de enero, por el que se aprueba la norma de calidad comercial para las patatas de consumo en el mercado nacional y se modifica el anexo I del Real Decreto 2192/1984, de 28 de noviembre, por el que se aprueba el Reglamento de aplicación de las normas de calidad para las frutas y hortalizas frescas comercializadas en el mercado interior. *BOE* **2009**, *21*, 8175–8182.

25. Becalski, A.; Lau, B.P.Y.; Lewis, D.; Seaman, S.W.; Hayward, S.; Sahagian, M.; Ramesh, M.; Leclerc, Y. Acrylamide in French fries: Influence of free amino acids and sugars. *J. Agric. Food Chem.* **2004**, *52*, 3801–3806. [CrossRef] [PubMed]

26. Williams, J.S.E. Influence of variety and processing conditions on acrylamide levels in fried potato crisps. *Food Chem.* **2005**, *90*, 875–881. [CrossRef]

27. Muttucumaru, N.; Powers, S.J.; Elmore, J.S.; Briddon, A.; Mottram, D.S.; Halford, N.G. Evidence for the complex relationship between free amino acid and sugar concentrations and acrylamide-forming potential in potato. *Ann. Appl. Biol.* **2014**, *164*, 286–300. [CrossRef] [PubMed]

28. De Wilde, T.; De Meulenaer, B.; Mestdagh, F.; Govaert, Y.; Vandeburie, S.; Ooghe, W.; Fraselle, S.; Demeulemeester, K.; Van Peteghem, C.; Calus, A.; et al. Influence of storage practices on acrylamide formation during potato frying. *J. Agric. Food Chem.* **2005**, *53*, 6550–6557. [CrossRef] [PubMed]

29. Burch, R.S.; Trzesicka, A.; Clarke, M.; Elmore, J.S.; Briddon, A.; Matthews, W.; Webber, N. The effects of low-temperature potato storage and washing and soaking pre-treatments on the acrylamide content of French fries. *J. Sci. Food Agric.* **2008**, *88*, 989–995. [CrossRef]

30. Abboudi, M.; Al-Bachir, M.; Koudsi, Y.; Jouhara, H. Combined effects of gamma irradiation and blanching process on acrylamide content in fried potato strips. *Int. J. Food Prop.* **2016**, *19*, 1447–1454. [CrossRef]

31. Kolek, E.; Simko, P.; Simon, P. Effect of NaCl on the decrease of acrylamide content in a heat-treated model food matrix. *J. Food Nutr. Res.* **2006**, *45*, 17–20.

32. He, F.J.; Tan, M.; Ma, Y.; MacGregor, G.A. Salt reduction to prevent hypertension and cardiovascular disease. *J. Am. Coll. Cardiol.* **2020**, *75*, 632–647. [CrossRef]

33. Matthäus, B.; Haase, N.; Vosmann, K. Factors affecting the concentration of acrylamide during deep-fat frying of potatoes. *Eur. J. Lipid Technol.* **2004**, *106*, 793–801. [CrossRef]

34. Mesias, M.; Delgado-Andrade, C.; Holgado, F.; Morales, F.J. Acrylamide content in French fries prepared in households: A pilot study in Spanish homes. *Food Chem.* **2018**, *260*, 44–52. [CrossRef]

35. Sayon-Orea, C.; Bes-Rastrollo, M.; Basterra-Gortari, F.J.; Beunza, J.J.; Guallar-Castillon, P.; de la Fuente-Arrillaga, C.; Martinez-Gonzalez, M.A. Consumption of fried foods and weight gain in a mediterranean cohort: The sun project. *Nutr. Metab. Cardiovas.* **2013**, *23*, 144–150. [CrossRef]

36. Guallar-Castillón, P.; Rodríguez-Artalejo, F.; Lopez-Garcia, E.; León-Muñoz, L.M.; Amiano, P.; Ardanaz, E.; Arriola, L.; Barricarte, A.; Buckland, G.; Chirlaque, M.D.; et al. Consumption of fried foods and risk of coronary heart disease: Spanish cohort of the European prospective investigation into cancer and nutrition study. *Br. Med. J.* **2012**, *344*, e363. [CrossRef] [PubMed]

37. Sánchez Muniz, F.J. Aceite de oliva, clave de vida en la Cuenca Mediterránea. *Real Acad. Nac. F.* **2007**, *73*, 653–692.

38. Fiselier, K.; Bazzocco, D.; Gama-Baumgartner, F.; Grob, K. Influence of the frying temperature on acrylamide formation in French fries. *Eur. Food Res. Technol.* **2006**, *222*, 414–419. [CrossRef]

39. Pedreschi, F.; Kaack, K.; Granby, K. Acrylamide content and color development in fried potato strips. *Food Res. Int.* **2006**, *39*, 40–46. [CrossRef]

40. Gökmen, V.; Senyuva, H.Z.; Dülek, B.; Çetin, A.E. Computer vision-based image analysis for the estimation of acrylamide concentrations of potato chips and French fries. *Food Chem.* **2007**, *101*, 791–798. [CrossRef]

41. Mesias, M.; Delgado-Andrade, C.; Holgado, F.; Morales, F.J. Impact of the consumer cooking practices on acrylamide formation during the preparation of French fries in Spanish households. *Food Addit. Contam. Part A* **2020**, *37*, 254–266. [CrossRef] [PubMed]

42. Nayak, P.K.; Dash, U.; Rayaguru, K.; Krishnan, K.R. Physio-chemical changes during repeated frying of cooked oil: A review. *J. Food Biochem.* **2016**, *40*, 371–390. [CrossRef]

Article

Assessment of Healthy and Harmful Maillard Reaction Products in a Novel Coffee Cascara Beverage: Melanoidins and Acrylamide

Amaia Iriondo-DeHond [1], **Ana Sofía Elizondo** [1], **Maite Iriondo-DeHond** [2], **Maria Belén Ríos** [1], **Romina Mufari** [1,3], **Jose A. Mendiola** [1], **Elena Ibañez** [1] and **Maria Dolores del Castillo** [1,*]

[1] Instituto de Investigación en Ciencias de la Alimentación (CIAL) (CSIC-UAM), Calle Nicolás Cabrera, 9, 28049 Madrid, Spain; amaia.iriondo@csic.es (A.I.-D.); anasofi.eliz@gmail.com (A.S.E.); rios.mbelen2019@gmail.com (M.B.R.); romi_mufari@hotmail.com (R.M.); j.mendiola@csic.es (J.A.M.); elena.ibanez@csic.es (E.I.)

[2] Instituto Madrileño de Investigación y Desarrollo Rural, Agrario y Alimentario (IMIDRA), N-II km 38, 28800 Alcalá de Henares, Spain; maite.iriondo@madrid.org

[3] Instituto de Ciencia y Tecnologia de los Alimentos (ICTA), Av. Velez Sarsfield 1611, Cordoba 5016, Argentina

* Correspondence: mdolores.delcastillo@csic.es; Tel.: +34-91-0017900 (ext. 953)

Received: 31 March 2020; Accepted: 7 May 2020; Published: 12 May 2020

Abstract: Our research aimed to evaluate the formation of Maillard reaction products in sun-dried coffee cascara and their impact on the safety and health promoting properties of a novel beverage called "Instant Cascara" (IC) derived from this coffee by-product. Maillard reaction products in sun-dried coffee cascara have never been reported. "Instant Cascara" (IC) extract was obtained by aqueous extraction and freeze-drying. Proteins, amino acids, lipids, fatty acid profile, sugars, fiber, minerals, and vitamins were analyzed for its nutritional characterization. Acrylamide and caffeine were used as chemical indicators of safety. Colored compounds, also called melanoidins, their stability under 40 °C and in light, and their in vitro antioxidant capacity were also studied. A safe instant beverage with antioxidant properties was obtained to which the following nutritional claims can be assigned: "low fat", "low sugar" "high fiber" and "source of potassium, magnesium and vitamin C". For the first time, cascara beverage color was attributed to the presence of antioxidant melanoidins (>10 kDa). IC is a potential sustainable alternative for instant coffee, with low caffeine and acrylamide levels and a healthy composition of nutrients and antioxidants.

Keywords: acrylamide; coffee cascara; food safety; instant beverage; Maillard reaction; melanoidins

1. Introduction

Maillard Reaction Products (MRPs) are the result of a chemical reaction between amino acids and reducing sugars when foods are being processed at high temperatures [1]. This reaction enhances flavor and color, and MRP have been associated with both positive and negative health effects [2]. In the case of coffee, the main MRPs produced during coffee roasting are melanoidins. The main sources of dietary melanoidins in Western diets are coffee and bakery products [3,4]. The average melanoidin content in medium roasted coffee is 7.2 g per 100 g. On the other hand, the amount of melanoidins in bread crusts ranges from 14 g to 30 g per 100 g of bread crust [3]. Different health-promoting properties such as antioxidant, antimicrobial, anti-inflammatory and antihypertensive activity have been assessed in these molecules [5]. Besides melanoidins, another MRP found in coffee beverages is acrylamide. This compound is considered a contaminant and classified by the International Agency of Research on Cancer (IARC) as a potential carcinogenic (class 2A) [6]. The European Food Safety Authority (EFSA) has stated that coffee and its substitutes can contribute up to 40% of the dietary exposure to acrylamide for the adult population. The European Commission (EC) established several recommendations for monitoring acrylamide levels in food, recommending 450 µg/kg for roasted coffee and 900 µg/kg for instant coffee [7]. Besides coffee, the main sources of human dietary exposure to acrylamide are fried potatoes (~272–570 µg/kg), bakery products (~75–1044 µg/kg) and breakfast cereals (~149 µg/kg) [7,8].

Given the great demand for coffee worldwide, an important number of by-products are generated during its processing [9]. Coffee cascara represents the main by-product of the coffee industry and its revalorization has gained interest over the last decade. After de-pulping, this by-product is normally dehydrated in the sun for 21 days to reduce its moisture to 10%. Cascara contains the substrates, such as amino acids, proteins and carbohydrates, needed for the Maillard reaction (MR) [10]. During the dehydration process of cascara, the MR may occur and healthy (melanoidins) and harmful (acrylamide) MRP may be generated [11,12]. The occurrence of the MR during the dehydration of figs, dates and raisins under similar conditions to those employed in cascara processing has been previously described [11].

Recent studies consider cascara as a potential source of phenolic compounds with antioxidant properties and the potential to improve human health [13,14]. For this reason, the use of this by-product to elaborate a novel antioxidant beverage supposes several advantages: promoting the sustainability of the coffee industry, avoiding the waste of new by-products by using the insoluble residue that results from the extraction to obtain cascara flour [10], and creating an added value, sustainable and healthy drink [15]. Since cascara is considered a novel food according to the European Commission [16], more safety and toxicity studies are needed for its further approval [17].

An aqueous extraction of coffee cascara was proposed to obtain a soluble powder also called hereby as 'Instant Cascara' (IC). This research seeks to obtain a product that offers a healthier nutritional profile than the powdered soft drinks commercially available, which stand out mainly for the excessive presence of sugar [18]. A very recently published critical review showed that the consumption of soft drinks has increased dramatically over the past years, being mostly consumed by children and teenagers [19]. Excessive intake of soft drinks with high sugar and acid leads to dental caries and erosion, overweight, obesity and increased risk of type 2 diabetes [19]. Although consumers are aware of the impact of the consumption of these beverages on human health, it is still necessary to educate the population about the harmful effects of these drinks and also to develop healthier alternatives. Therefore, the aim of this research was to develop a novel safe antioxidant "Instant Cascara" beverage (IC), contributing to the sustainability of the coffee sector and offering healthier products to the general population satisfying their nutritional demands. To achieve the goal, the formation of Maillard reaction products in sun-dried coffee cascara and their impact on the safety and health promoting properties of a novel beverage called "Instant Cascara" (IC) derived from this coffee by-product was assessed. Maillard reaction products in sun-dried coffee cascara have never been reported before.

2. Materials and Methods

2.1. Food Samples

2.1.1. Raw Cascara

SUPRACAFÉ S.A. (Móstoles, Madrid, Spain) provided coffee cascara (CA) from Arabica species and Tabi variety from Colombia. Coffee cascara was obtained in the processing of the coffee berry, dried for 21 days in the sun, and subjected to a sanitation process involving the use of a carbon dioxide atmosphere (Martin Bauer, MABA-PEX process).

2.1.2. Homemade Instant Cascara

Powdered aqueous extract from coffee cascara (IC) was obtained as described in the patent WO2013004873A1 [20], which consisted of an aqueous extraction of 50 g/L at 100 °C for 10 min. Sample was filtered (250 µm) and freeze-dried. IC extraction yield was 20%.

Two beverages were formulated from IC at 4 mg/mL and 10 mg/mL. Concentrations were chosen based on commercial instant coffee drinks. Elaboration consisted of diluting each IC dose with water at room temperature.

2.1.3. Commercial Cascara Infusion (Tabifruit)

IC drinks (4 and 10 mg/mL) were compared with a commercial infusion of coffee cascara (Tabifruit, Supracafé S.A., Madrid, Spain). This was elaborated according to the procedure indicated for the product, leaving the infusion bag (3 g) in 250 mL of water at 100 °C for 4 min, creating a final concentration of cascara of 12 mg/mL.

2.2. Nutritional Characterization

2.2.1. Protein and Amino Acid Profile

Protein content in cascara (CA) and instant cascara (IC) was obtained through Kjeldahl mineralization carried out by the Bioanalytical Techniques Unit at the Instituto de Investigación en Ciencias de la Alimentación (CIAL, UAM-CSIC, Madrid, Spain) and quantification was performed by colorimetric analysis of nitrogen (AOAC-32.1.22,920.87). A conversion factor of 5.6 was used to calculate protein content. Analysis was carried out in duplicate and results were expressed as percentage of dry matter.

Free and total amino acid quantification was carried out by the Servicio de Química de Proteínas of the Centro de Investigaciones Biológicas (CIB, CSIC, Madrid, Spain). For total amino acid quantification, samples were hydrolyzed in an acid medium for later analysis by high performance liquid chromatography (HPLC) with post-column derivatization, using nihydrin. Analyses were carried out in duplicate and results were expressed as mg/g.

Both determinations were performed as previously described [21].

2.2.2. Lipids and Fatty Acid Profile

Lipid quantification in CA and IC was performed by Soxhlet extraction as described in AOAC Official Method 945.16. using petroleum ether. Results were obtained by weighing the dry product containing the lipid fraction and were expressed as % dry matter.

Fatty acid profile was determined by gas chromatography according to ISO 12966-2:2017; using a flame ionization detector (Agilent 7820A GC system, Agilent Technologies, Inc., Santa Clara, CA, USA) [22]. Analysis was carried out by the Analysis Services Unit facilities of the Institute of Food Science, Technology and Nutrition (ICTAN, CSIC, Madrid, Spain). Analysis was carried out in duplicate and results were expressed as grams per 100 g of sample.

Both determinations were performed as previously described [23].

2.2.3. Dietary Fiber

Insoluble (IDF), soluble (SDF) and total (TDF) dietary fiber were determined in CA and IC using the Megazyme Total Dietary Fiber Kit (Megazyme, Wicklow, Ireland), an enzymatic-gravimetric assay based on the AOAC-991.43 and AACC-32.07.01 method. Analysis was carried out in duplicate. Results were expressed as weight percentage (%).

2.2.4. Sugars, Minerals and Vitamin C

Determination of sugars, minerals and ascorbic acid in CA and IC was carried out by the Analysis Services Unit facilities of the Institute of Food Science, Technology and Nutrition (ICTAN, CSIC, Madrid, Spain). All determinations were performed in duplicate as previously described [24]. Results were expressed as g/100 g for simple sugars and mg/100 g for minerals and ascorbic acid.

2.2.5. Total Carbohydrates

Carbohydrate content in liquid IC beverages (4 mg/mL and 10 mg/mL) was determined as described by Masuko et al. (2005), using the phenol-sulfuric method [25]. The experiment was initiated with the preparation of the reagents, phenol at 5% (Sigma-Aldrich, St. Louis, MO, USA) and sulfuric acid at 98% (Sigma-Aldrich, St. Louis, MO, USA). Glucose (Sigma-Aldrich, St. Louis, MO, USA) calibration curves was prepared (0.1–0.5 mg/mL), and in 2 mL glass vials, 100 µL of sample or standard, 300 µL of sulfuric acid and 90 µL of phenol or H_2O for the sample blanks were added. Vials were incubated at 90 °C for 5 min, followed by a water bath at room temperature for 5 min. Then, 100 µL of each sample was transferred to a 96-well plate and absorbance was measured at 490 nm in a UV-Visible spectrophotometer (BioTek Instruments, Winooski, VT, USA).

2.2.6. Glucose

Free glucose content was determined in liquid IC beverages (4 mg/mL and 10 mg/mL) using the Glucose TR Kit (SpinReact, Girona, Spain) following manufacturer's instructions. Results were expressed as mg glucose/ mL of beverage.

2.3. Maillard Reaction Products (MRP)

2.3.1. Acrylamide

For certifying the safety of the novel instant beverage, acrylamide content was analyzed in the IC powder and in liquid samples. A sample preparation procedure according to the standard UNE-EN 16618:2015 [26] with some modifications was used. A total of 1 g of sample was weighed into a centrifuge tube (20 mL of water in the case of IC powder were added) and was spiked with 100 ng/g of $^{13}C_3$-labelled acrylamide (Sigma-Aldrich, St. Louis, MO, USA), to determine the percentage recovery of the method at this stage. Then, it was quantitatively transferred for the SPE clean-up in C18 (Isolute, 500 mg/ 6 mL, Biotage, Uppsala, Sweden) cartridge preconditioned using a multi-stage vacuum system (Agilent Technologies; Palo Alto, CA, USA). The eluate was collected and submitted to another step of clean up with the ENV+ (Isolute, 500 mg/ 6 mL, Biotage, Uppsala, Sweden) preconditioned cartridge. The cartridge was rinsed with two fractions of 2 mL of water, followed by the elution of acrylamide with 2 mL of methanol 60% in water. The final extract was concentrated by removing the extraction solvent in an oven at 35 °C to an approximate volume of 500 µL and stored in vials for the quantitation by LC-MS/MS.

The LC-MS/MS analysis was performed according to methodology described by Mastovska and Lehotay [27], using an Agilent 1290 Infinity II system, interfaced to an Agilent 6470 triple quadrupole mass spectrometer (Agilent Technologies; Palo Alto, CA, USA). Sample injection volume was 10 µL, and an Agilent Infinity Lab Poroshell 120 EC-C18 column (100 × 4.6 mm; 2.7 µm particle size) kept at 30 °C was employed for the LC separation. The mobile phase was 99.5:0.5 (*v/v*) water-MeOH, with 0.1% formic acid, at a flow rate of 400 µL/min for 8 min for elution of acrylamide (retention time 4.02 min) and then 0.1% formic acid in MeCN-MeOH (50:50, *v/v*) for the post-analysis wash (at 400 µL/min for 6 min) followed by equilibration to initial conditions. The MS determination was performed in ESI positive mode. Monitoring transitions m/z 72 > 55, 72 > 54, 72 > 44, and 72 > 27 were recorded for acrylamide, whereas the transitions m/z 75 > 58 were used for the $^{13}C_3$-labelled acrylamide. Results were expressed as µg/kg of sample.

2.3.2. Melanoidins

The content of melanoidins in liquid IC beverages was analyzed spectrophotometrically. Light-absorption measurement of samples at 420 nm was performed using a microplate reader (BioTek Epoch 2 Microplate Spectrophotometer, Winooski, VT, USA). Caramel (E-150d) was used as a melanoidin standard. Analytical determination was carried out in triplicate. Results were expressed in equivalent milligrams of caramel melanoidins/gram of sample.

2.3.3. Antioxidants

Preliminary Information on Antioxidant Composition by Spectral Analysis

UV-Visible spectrum of in liquid IC beverages (4 mg/mL and 10 mg/mL) was obtained using a microplate reader (BioTek Epoch 2 Microplate Spectrophotometer, Winooski, VT, USA). Absorption spectra of samples were recorded from 200 to 700 nm.

Phenolic Compounds

For phenolic compounds analysis in liquid IC beverages (4 mg/mL and 10 mg/mL), Folin-Ciocalteu method was adapted to a micro-method format [28]. Folin-Ciocalteu reagent (Sigma-Aldrich, St. Louis, MO, USA) and chlorogenic acid (CGA) (Sigma-Aldrich, St. Louis, MO, USA) standard solution (0.1–0.9 mg/mL) were prepared. Reaction was initiated by adding 10 μL of sample or standard and 150 μL of the Folin solution to a 96-well plate. Blanks from samples and reagent were also analyzed. After 3 min of incubation at 37 °C, 50 μL of sodium carbonate (Sigma-Aldrich, St. Louis, MO, USA) were added. Samples were incubated for 2 h and absorbance was measured at 735 nm. Results were expressed as mg/mL of CGA and measurements were performed in triplicate.

Total Anthocyanins

Total anthocyanins content was measured in liquid IC beverages (4 mg/mL and 10 mg/mL) according to Wrolstad and Giusti (2001) [29]. In a 96-well plate, 40 μL of sample and 160 μL of potassium chloride (0.025 M, Sigma-Aldrich, St. Louis, MO, USA) and sodium acetate buffer (0.4 M, Sigma-Aldrich, St. Louis, MO, USA) were added and incubated for 15 min at 37 °C. Absorbance was measured at 520 nm and 700 nm. Cyanidine-3-glucoside (C3G) was used as reference for calibration curve (0.0–0.2 mg/mL) and measurements were performed in triplicate.

Overall Antioxidant Capacity

ABTS

ABTS (2,2′-azino-bis(3-ethylbenzothiazoline-6-sulfonic acid)) bleaching method was determined in liquid IC beverages (4 mg/mL and 10 mg/mL) as detailed by Tsao et al. (2003) and rectified by Oki et al. (2006) for its use in a microplate [30,31]. ABTS$^{\bullet+}$ stock solution was prepared by mixing the ABTS$^{\bullet+}$ radical and potassium persulfate (Sigma-Aldrich, St. Louis, MO, USA). Solution was then left to stand for 16h at room temperature. Afterwards, ABTS$^{\bullet+}$ working solution was prepared by diluting the stock solution 1:75 (*v/v*) in 5 mM of sodium phosphate buffer at pH 7.4 and adjusted to an absorbance of 0.7 ± 0.02 at 734 nm. Chlorogenic acid (CGA) calibration curve (25–200 μM) was used for antioxidant capacity analysis. Measurements were performed in triplicate and results were expressed as mg/mL of CGA.

FRAP

Antioxidant capacity by FRAP (Ferric Reducing Antioxidant Power Assay) was determined in liquid IC beverages (4 mg/mL and 10 mg/mL) as described by Benzie and Strain (1996) and modified by Tsao et al. (2003) for use in a microplate [30,32]. Experiment was initiated with the preparation of TPTZ 10 mM in 40 mM HCl, ferric chloride hexahydrate ($FeCl_3 \cdot 6H_2O$) 20 mM, and the FRAP reagent (Sigma-Aldrich, St. Louis, MO, USA): a mixture of 0.3 M of acetate buffer, 2.5 mL of TPTZ solution and 2.5 mL of $FeCl_3 \cdot 6H_2O$ solution. Reaction took place by adding 10 μL of sample and 290 μL of FRAP reagent into a 96-well plate for 10 min at 593 nm. Sample and reagent blanks were also measured. CGA was used for the calibration curve (0.025–0.2 mg/mL) and results were expressed as mg/mL of CGA.

Contribution of Melanoidins to the Overall Antioxidant Properties of IC

To determine which compounds contribute to the antioxidant properties of the beverages, each sample was ultra-filtrated using Macrosep Advance Centrifugal Devices (Pall Corporation, Ann Arbor, MI, USA) of a molecular cut membrane of 10 kDa. Samples were centrifuged (Hettich Universal 320R Centrifuge, Andreas Hettich GmbH & Co.KG, Tuttlingen, Germany) at 3000 g for 90 min to separate the high molecular weight (HMW) and low molecular weight (LMW) fractions. HMW fractions were washed three times using same centrifugation conditions. Antioxidant capacity of HMW and LMW fractions of IC beverages (4 and 10 mg/mL) and Tabifruit was analyzed by ABTS and FRAP methods (see methods described above).

2.4. Shelf Life Study Under Accelerated Storage Conditions of Liquid IC

Beverage stability was evaluated under different light exposure and temperature conditions. In 10 mL glass vials, samples were tested for 72 h under three different conditions: 40 °C and light (Temp + light), room temperature and light (light) and room temperature and no light exposure (darkness) [33]. Light exposure was carried out with artificial light from a fluorescent lamp of 500 lux of illuminance. Prior to the study, pH of each sample was measured using the pH meter MP 230 (Mettler Toledo, Barcelona, Spain) previously calibrated.

Total phenolic content and ABTS were determined in beverages after the shelf life study as indicated in Section 2.3.3.

Color

Color measurement was performed with a Reflectance Integrating Sphere SPECORD 210 Plus (Analytik Jena, Jena, Germany). Samples were analyzed in 2 mL plastic vials and results were obtained through the WinAspect Plus program (Jena, Germany), using the CIE color space L* a* b* as numerical values representing luminosity and color parameters. Samples were analyzed in triplicate. Colors were generated from L* a* b* values using colorizer.org as a high precision color generator [34].

2.5. Statistical Analysis

All results were expressed as the mean ± standard deviation (SD). Analysis of variance (ANOVA) and Tukey as a post-hoc test were carried out to determine differences between means. Differences were considered to be statistically significant at $p < 0.05$. Pearson's correlation coefficient was calculated using the Excel Analysis *ToolPak* (Microsoft, Redmond, WA, USA). XL-Stat version 2020.1.3 (Addinsoft, New York, NY, USA) was used to analyze JAR data. Analysis of variance were conducted on IBM SPSS Statistics 24 (IBM, Armonk, NY, USA).

3. Results and Discussion

3.1. Nutritional Characterization of CA

Table 1 shows protein and amino acid content (free and total) of the raw material used in this investigation, coffee cascara (CA). Protein content found in CA was 9.55%, which is in accordance to that previously described [10,35–38].

Table 1. Total protein (%) and free and total amino acid content (mg/g) of raw dry coffee cascara (CA) and powdered Instant Cascara beverage (IC).

	CA		IC	
Total Protein (%)	**9.55 ± 0.11** [b]		**6.25 ± 0.27** [a]	
Amino Acids (mg/g)	**Free**	**Total**	**Free**	**Total**
Aminobutyric acid (GABA)	N.D.	N.A.	0.3 ± 0.00	N.A.
Glutamic acid (Glu)	N.D.	2.13 ± 0.04 [a]	0.40 ± 0.01	2.11 ± 0.17 [a]
Alanine (Ala)	N.D.	2.06 ± 0.11 [a]	2.27 ± 0.01	2.44 ± 0.05 [a]
Arginine (Arg)	N.D.	0.48 ± 0.03 [a]	1.39 ± 0.02	1.05 ± 0.04 [b]
Asparagine (Asp)	N.D.	2.84 ± 0.14 [a]	1.91 ± 0.05	7.79 ± 0.37 [b]
Cysteine (Cis)	N.D.	0.18 ± 0.00 [a]	0.17 ± 0.01	0.22 ± 0.02 [a]
Phenylalanine (Phe)	N.D.	0.98 ± 0.10 [a]	0.17 ± 0.01	0.31 ± 0.01 [a]
Glycine (Gly)	N.D.	2.71 ± 0.17 [a]	0.30 ± 0.27	1.19 ± 0.04 [a]
Histidine (His)	N.D.	0.52 ± 0.10 [b]	0.08 ± 0.01	0.16 ± 0.01 [a]
Isoleucine (Ile)	N.D.	0.63 ± 0.05 [b]	0.05 ± 0.00	0.18 ± 0.04 [a]
Leucine (Leu)	N.D.	1.17 ± 0.03 [b]	0.05 ± 0.01	0.27 ± 0.04 [a]
Lysine (Lys)	N.D.	0.38 ± 0.01 [a]	0.03 ± 0.01	0.21 ± 0.01 [a]
Methionine (Met)	N.D.	0.25 ± 0.02 [a]	N.D.	0.07 ± 0.04 [a]
Proline (Pro)	N.D.	1.60 ± 0.01 [a]	6.77 ± 0.02	4.82 ± 0.12 [b]
Serine (Ser)	N.D.	1.91 ± 0.02 [a]	5.71 ± 0.01	2.08 ± 0.05 [a]
Tyrosine (Tyr)	N.D.	0.63 ± 0.05 [a]	N.D.	0.68 ± 0.01 [a]
Tryptophan (Trp)	N.D.	N.A.	N.D.	N.A.
Threonine (Thr)	N.D.	0.96 ± 0.03 [b]	0.05 ± 0.01	0.35 ± 0.02 [a]
Valine (Val)	N.D.	1.27 ± 0.02 [a]	0.30 ± 0.01	0.60 ± 0.07 [a]
Totals (mg/g)	N.D.	25.05 ± 0.16 [a]	20.43 ± 0.37	29.78 ± 1.97 [b]
EAA (% total)	N.D.	32.07 ± 0.07 [b]	10.19 ± 0.01	13.04 ± 0.25 [a]
BCAA (Val + Leu + Ile) (% total)	N.D.	14.83 ± 0.10 [b]	1.92 ± 0.01	4.28 ± 0.14 [a]
AAA (Phe + Tyr + Trp) (% total)	N.D.	7.77 ± 0.15 [b]	0.8 ± 0.01	4.03 ± 0.00 [a]

EAA, essential amino acids; BCAA, branched-chain amino acids; AAA, aromatic amino acids. N.D., not detected; N.A., not analyzed. Results are expressed as mean ± SD. Different superscript letters indicate significant differences for total protein and total amino acids (Student's T, $p < 0.05$).

With regard to amino acid content, no free amino acids were detected in CA. Amino acids present in the raw material derived from the proteins that compose it. Of the total amino acids in CA, 32% corresponded to essential amino acids. Asparagine, glycine and glutamic acid presented the highest values for total amino acids in CA. The main amino acids found in CA corresponded to those reported by Elías in 1979 [39] and also recently described by our research group [10]. Tryptophan was not detected in total amino acids, as the acidic conditions needed for quantification resulted in its hydrolysis.

The lipid content and the fatty acid profile of CA are shown in Table 2. Total fat content found in CA was 2%, which is in line to that reported by other authors (2.3–2.5%) [38,40,41]. In this study, palmitic acid (C16:0) was the main fatty acid in CA (36.02 g/100 g), followed by linoleic acid (C18:2n6c, 21.80 g/100 g), α-linoleic acid (C18:3n3, 17.37 g/100 g) and oleic acid (C18:1n9c, 6.72 g/100 g). The polyunsaturated fatty acids (PUFA) and saturated fatty acids (SFA) ratio was 0.85, i.e., higher than 0.45, which is considered as healthy by the World Health Organization [42].

The main sugars found in CA were glucose (13.45 g/100 g) and fructose (21.70 g/100 g). Mannose (0.06 g/100 g) was also detected in CA. Xylose and sucrose were not detected in this by-product. Values for simple sugars are in line with that reported by Urbaneja et al. (1996) [43].

With regard to fiber content in CA, this by-product presented 47.44% of total dietary fiber (TDF). This dietary fiber is composed of fibers of different nature, being 31.32% insoluble (IDF) and 16.12% soluble (SDF). These values are in line with those reported for CA in other studies and by companies that are using this by-product as a food ingredient [10,44].

Table 2. Total fat content (%) and fatty acid profile (g/100g) of raw dry coffee cascara (CA) and powdered Instant Cascara beverage (IC).

	CA	IC
Total Fat (%)	2.00 ± 0.50 [b]	0.58 ± 0.18 [a]
Fatty Acid Profile (g/100 g)		
C12:0	0.10 ± 0.02 [a]	N.D [a]
C14:0	1.18 ± 0.02 [a]	5.77 ± 0.41 [b]
C15:0	0.37 ± 0.01 [a]	4.02 ± 0.59 [a]
C16:0	36.02 ± 0.32 [a]	30.54 ± 3.27 [b]
C16:1n7	3.04 ± 0.19 [a]	3.90 ± 0.23 [a]
C17:0	0.56 ± 0.01 [a]	N.D [b]
C18:0	5.64 ± 0.22 [a]	4.54 ± 0.37 [a]
C18:1n7c	1.79 ± 0.05 [a]	N.D [b]
C18:1n9c	6.72 ± 0.37 [a]	10.82 ± 1.80 [a]
C18:2n6c	21.80 ± 0.34 [a]	15.83 ± 1.00 [a]
C18:3n3	17.37 ± 0.24 [a]	14.84 ± 1.46 [a]
C20:0	2.82 ± 0.07 [a]	N.D [a]
C20:1n9	0.09 ± 0.00 [a]	N.D [a]
C20:2n6	0.09 ± 0.00 [a]	N.D [a]
C20:3n3	0.19 ± 0.04 [a]	N.D [a]
C20:5n3	N.D [a]	N.D [a]
C21:0/C20:3n6 *	0.10 ± 0.01 [a]	N.D [a]
C22:0	0.64 ± 0.04 [a]	N.D [b]
C22:6n3	0.76 ± 0.60 [a]	9.74 ± 5.54 [a]
C23:0	0.16 ± 0.02 [a]	N.D [a]
C24:0/C22:5n3 *	0.57 ± 0.00 [a]	N.D [b]
SFA (%)	47.48 ± 0.10 [a]	44.87 ± 4.65 [a]
MUFA (%)	11.64 ± 0.12 [a]	14.72 ± 1.57 [a]
PUFA (%)	40.88 ± 0.02 [a]	40.41 ± 3.08 [a]

SFA, saturated fatty acids; MUFA, monounsaturated fatty acids; PUFA, polyunsaturated fatty acids; N.D., not detected. The values indicate the mean ± SD. Different superscript letters indicate significant differences (Student's T. $p < 0.05$). * The chromatographic method does not allow the separation of the fatty acids C21:0 and C20:3n6; and C24:0 and C22:5n3, so the value obtained may be due to either one or the sum of both.

Considering micronutrients present in CA, potassium (2284 mg/100 g), magnesium (20.84 mg/100 g), sodium (266.58 mg/100 g) and calcium (54.78 mg/100 g) were the cations detected in this by-product. As for anions, chloride (473.21 mg/100 g), nitrate (127.14 mg/100 g), phosphate (602.79 mg/100 g) and sulfate (193.32 mg/100 g) were also found in CA. Micronutrient values reported in this study are slightly higher than those previously described [39]. In addition, CA presented 69.70 mg/100 g of ascorbic acid. Content of ascorbic acid reported in this study is much higher than that described by The Coffee Cherry Co. for their cascara flour [44]. This difference may be due to many factors influencing the nutritional composition of CA, such as origin, cultivation methods and variety of the coffee plant, among others [45].

3.2. Characterization of IC

3.2.1. Nutritional Profile

After an aqueous extraction and freeze-drying process, a powdered extract (IC) was obtained from CA. A nutritional characterization was also carried out in the novel instant beverage. With regard to macronutrients, 65% of the protein present in CA is recovered in IC (6.25%). To the best of our knowledge, this is the first time that the amino acid profile of an aqueous extract of coffee cascara is described (Table 1). IC presented 25 mg/g of free amino acids, 10.19% of them being essential. The aqueous extraction process resulted in a concentration of free amino acids that were not present in

CA. From a nutritional point of view, free amino acids would be more bio-accessible in IC compared to CA. Proline (6.77 mg/g), serine (5.71 mg/g) and alanine (2.27 mg/g) were the major free amino acids found in IC. In addition, free aminobutyric acid (GABA) was also detected in IC (0.3 mg/g). GABA is known for its health-promoting properties, such as an anti-hypotensive effect [46]. Together with other fruits, such as melon and apricot, IC could contribute to lower blood pressure. Considering results obtained in free and total amino acids, IC might be a potential sustainable source of these molecules. A recent clinical trial has observed that continuous intake of the amino acid supplements significantly increase muscle amount and improve skin texture in young adult women [47].

With regard to total amino acids, asparagine was the main amino acid in IC (7.79 mg/g), followed by proline (4.82 mg/g) and alanine (2.44 mg/g). Although the claimed health-promoting properties of amino acids are not yet established in terms of a cause–effect relationship after the evaluation of the EFSA Panel, the potential claimed health-promoting properties related to amino acids are growth or maintenance of muscle mass, maintenance of normal muscle function, faster recovery of muscle function/strength/glycogen stores after exercise, faster recovery from muscle fatigue after exercise and skeletal muscle tissue repair [48].

Total fat percentage is significantly higher ($p < 0.05$) in CA (2%) than in IC (0.58%). According to the statement by the European Commission Regulation No 1924/2006, IC would be "low in fat" since the product contains no more than 1.5 g of fat per 100 mL [49]. This instant beverage would even be close to the "fat-free" nutrition claim, which is attributed to products that have no more than 0.5 g of fat per 100 mL. Fatty acid composition of IC mainly includes palmitic acid (C16:0), followed by linoleic acid (C18:2n6) and α-linoleic acid (C18:3n3), similar to that found in CA (Table 2).

The main simple sugars found in IC were fructose (16.19 g/100g) and glucose (6.02 g/100 g), being lower in IC compared to CA. Xylose (6.02 g/100 g), sucrose (0.08 g/100 g) and mannose (0.03 g/100 g) were also detected in IC.

Dietary fiber present in IC was 18.32%, all soluble dietary fiber. As expected, the aqueous extraction process concentrated the SDF present in CA. There are several health promoting properties attributed to SDF, which include reduction in cholesterol level and blood pressure, prevention of gastrointestinal diseases, protection against onset of several cancers, such as colorectal, prostate and breast cancer, and increased mineral bioavailability, among others [50,51]. This novel powdered beverage can reach the nutrition claim of "high in fiber" since the product contains at least 6 g of fiber per 100 g [49]. The health claims attributed to the "high in fiber" nutrition claim are "fiber increases fecal bulk, contributes to normal bowel function and to an acceleration of intestinal transit" [52].

After aqueous extraction, powdered IC was enriched in micronutrients since values of anions and cations in IC were higher compared to CA. Cations in IC were potassium (6701 mg/100 g), magnesium (121.56 mg/100 g), sodium (354.19 mg/100 g) and calcium (109.88 mg/100 g). As for anions, chloride (618.32 mg/100 g), nitrate (489.48 mg/100 g), phosphate (1314 mg/100 g) and sulfate (533.61 mg/100 g) were detected in IC.

The European Commission (EU) regulation N° 1925/2006, indicates that to establish an ingredient as a source of any micronutrient, it must represent at least 15% of the daily recommendation [53]. Recommended daily allowances for potassium and magnesium are 3600 and 300 mg, respectively. Therefore, IC may be considered a "source of potassium and magnesium" since values of potassium and magnesium present in IC represent 15% of the recommended allowance per 100 g of product. The Official Journal of the European Union (No 432/2012) makes the following statement for foods considered as a source of potassium: "Potassium contributes to the normal functioning of the nervous system, muscles, and the maintenance of normal blood pressure" [52]. On the other hand, a product that is a "source of magnesium" is related to the following health claims: "Magnesium contributes to a reduction of tiredness and fatigue, to electrolyte balance, to normal energy-yielding metabolism, to normal functioning of the nervous system, to normal muscle function, to normal protein synthesis, to normal physiological function, to the maintenance of normal bones and teeth and a role in the process of cell division" [52].

With regard to the content of ascorbic acid, IC presented 438.95 mg/100 g. IC can also be considered as a source of vitamin C, considering that the daily recommendation is 60 mg/day. Thus, the following health claims can be attributed to IC: "Vitamin C contributes to the normal functioning of the immune system during intense physical exercise, normal energy metabolism, normal functioning of the nervous system, normal psychological function, protection of cells against oxidative damage, reduction of fatigue and fatigue, regeneration of the reduced form of vitamin E, improvement of iron absorption, and normal collagen formation for the normal functioning of blood vessels, bones, cartilage, gums, skin, and teeth" [52].

Once powdered IC was characterized, two liquid beverages were prepared at 4 and 10 mg/mL. Results of liquid beverages were compared to the commercial cascara infusion Tabifruit. Table 3 shows the composition in nutrients and antioxidants of the three drinks. All samples showed significant differences ($p < 0.05$) in carbohydrate content. Tabifruit presented the lowest values of total carbohydrates (22 g/100 mL) followed by IC at 4 mg/mL (27 g/100 mL) and IC at 10 mg/mL (47 g/100 mL). This carbohydrate fraction may be composed of the soluble dietary fiber previously mentioned and other polysaccharides. For instance, previous studies showed that coffee cascara contains up to 35% of pectin [54].

Table 3. Nutrients (total carbohydrates (g/100 mL), glucose content (g/100 mL)) and non-nutrient antioxidants of the three beverages IC (4 mg/mL), IC (10 mg/mL) and Tabifruit.

	IC (4 mg/mL)	IC (10 mg/mL)	Tabifruit
Nutrients			
Total carbohydrates (g/100 mL)	27.50 ± 0.07 [b]	47.48 ± 0.20 [c]	22.22 ± 0.06 [a]
Glucose (g/100 mL)	0.04 ± 0.01 [b]	0.05 ± 0.01 [c]	0.02 ± 0.01 [a]
Antioxidants			
Total phenolic content (mg eq. CGA/mL)	0.25 ± 0.0 [a]	0.89 ± 0.0 [c]	0.37 ± 0.0 [b]
Anthocyanins	N.D.	N.D.	N.D.
ABTS (mg eq. CGA/mL)	16.82 ± 0.9 [b]	27.58 ± 2.3 [c]	15.05 ± 0.6 [a]
FRAP (mg eq. CGA/mL)	0.15 ± 0.0 [a]	0.43 ± 0.0 [c]	0.34 ± 0.0 [b]

CGA, chlorogenic acid. Each value represents the mean ± SD. Different letters indicate significant differences ($p < 0.05$) between samples in the same row. Different superscript letters indicate significant differences between samples (Tukey Test, $p < 0.05$).

Regarding glucose content, the same behavior as for total carbohydrates was observed; Tabifruit showed the lowest values (0.02 g/100 mL) and IC at 10 mg/mL the highest (0.05 g/100 mL). As expected, a dose-response effect in total carbohydrates and glucose content was observed in IC at 4 and 10 mg/mL. For both IC beverages, glucose values remained under those established for the nutrition claim "low in sugar" and could even be classed as "sugar-free". According to the European Commission Regulation (No. 1047/2012), for a food to be considered "low sugar", it must contain no more than 2.5 g of sugar per 100 mL; and for it to be declared "sugar-free", the product must not contain more than 0.5 g of sugar per 100 mL [49]. Therefore, considering free glucose content both IC beverages could make the "low sugar" nutrition claim. For the "sugar-free" claim, further quantification of other predominant simple sugars would be necessary. IC is a healthy alternative compared to other instant powdered soft drinks in the Food Data Central Database of the United States Department of Agriculture (USDA), whose sugar content ranges from 29 to 93 g/100 g per product [55].

3.2.2. Impact of MRP on Safety and Health Promoting Properties of IC

Safety

MRP compromise the nutritional value, safety and health promoting properties of IC. Since IC contains amino acids (Table 1) and reducing sugars, which are substrates of the Maillard reaction, acrylamide was analyzed to confirm the food safety of the novel beverages. Acrylamide content was below the detection limit for the three beverages. In contrast, it was found in powdered IC at 223 µg/kg. Acrylamide content found in IC is much lower than the limit stablished by the European Commission for coffee and instant coffee, 450 and 900 µg/kg, respectively [7]. Considering acrylamide content, this novel instant beverage would be a safe alternative to instant coffee. However, since acrylamide compromises the food safety of the beverage, shorter drying periods, different drying techniques or using fresh coffee cascara are measures that must be considered to decrease the amount of this compound in IC.

Since asparagine was the main amino acid in IC and this amino acid is known to be responsible for the development of acrylamide, another potential acrylamide mitigation strategy would be treating IC with L-asparaginase [56]. This enzyme is considered to be useful for acrylamide mitigation and to have negligible effects on the general formation of Maillard products. L-Asparaginase can selectively reduce the level of free L-Asn by hydrolyzing it to L-Asp and ammonia, thus specifically removing one of the essential acrylamide precursors [57]. However, in this particular case the preferred acrylamide mitigation strategy would be non-thermal drying procedures of cascara.

Drying of the raw material is a critical step in the conversion of coffee cascara into a safe food ingredient for human consumption. Novel drying methods will ensure the chemical and microbiological safety of cascara. Reducing moisture under 13% is necessary to avoid fungal growth and the consequent production of mycotoxins [58]. In this context, previous research has confirmed the absence of mycotoxins, such as aflatoxin B1, enniantin B and ochratoxin A in a cascara aqueous extract [59].

Another limitation on the use of a coffee by-product as a food ingredient is connected to its caffeine content. Results published so far suggest that caffeine content in an aqueous cascara extract (1.39%) does not need to be considered a safety concern [59]. The caffeine content of IC at 10 mg/mL in 250 mL would be around 34 mg, which is below the EFSA safety level for daily caffeine consumption of 400 mg for the general population, 3 mg/kg b.w. per day for children and adolescents and 200 mg for lactating women [60]. Therefore, almost no limitations on the use of IC for human consumption need to be considered, since pregnant women would have to drink over 1.47 L of IC beverage to exceed the safety level for the fetus.

Antioxidant Properties

Overall Antioxidant Capacity and Identification of Antioxidant Compounds

Liquid IC beverages are also a source of antioxidants (Table 3). In both antioxidant capacity determinations (ABTS and FRAP), IC at 10 mg/mL presented a significantly higher ($p < 0.05$) antioxidant capacity (27.58 mg eq. CGA/mL for ABTS and 0.43 mg eq. CGA/mL for FRAP) compared to the other two beverages. Results obtained for the antioxidant capacity corresponded to the total phenolic content (TPC) found in samples, IC being at 10 mg/mL the beverage presenting the highest values for TPC (0.89 mg eq. CGA/100 mL). Pearson's correlation coefficient was calculated to check for the existence of linear relationships between TPC, ABTS and FRAP of the beverages. A very strong (r = 0.97) or strong (r = 0.79) positive association was found between TPC and ABTS and FRAP, respectively. TPC, but not anthocyanins that were not detected in samples, seem to contribute to the overall antioxidant capacity of IC. Previous studies have reported that coffee cascara is a source of phenolic compounds, such as CGA and protocatechuic acid, which represent more than 80% of the polyphenols analyzed by Heeger et al. (2017) [40].

The presence of anthocyanins in coffee cascara has been previously reported [14,61,62]. However, anthocyanins were not detected in any of the samples using the pH-dependent colorimetric assay (Table 3) and therefore results seem to indicate that they are not the main responsible compounds for the antioxidant properties of IC. A comparative study of fresh and sun-dried grapes indicated a total loss of flavonoids, and in less quantity, of hydroxycinnamic acids (estimated at 62% for sun-dried grapes). This study concluded that the most labile polyphenols were procyanidins and flavan-3-ols, since they were completely degraded in all sun-dried raisin samples [63]. Therefore, anthocyanins might have been degraded during the 21 days of sun drying of the coffee cascara used in the present study, which, in fact, is the most cheap and common method to dry it. However, quantification analysis, such as NMR [64] and HPLC [65], are needed to confirm the absence of anthocyanins in the studied sample. In addition, the three beverages showed a similar UV-Visible absorption spectrum (Figure S1A). A maximum absorption was detected at 280 nm, indicative of the presence of caffeine, proteins and phenolic compounds, and at 325 nm, which would correspond to CGA and caffeic acid [66]. This analysis would also support the absence of anthocyanins that absorb at 520 nm [29] (Figure S1B), since none of the beverages seemed to absorb at this wavelength (Figure S1A).

With regard to Maillard reaction products, melanoidins were detected in the three beverages at 1.48 mg eq. caramel/mL, 0.54 mg eq. caramel/mL, 1.2 mg eq. caramel/mL, for IC (10 mg/mL), IC (4 mg/mL and Tabifruit, respectively. To the best of our knowledge, this is the first time the presence of melanoidins and the occurrence of the Maillard reaction are described in coffee cascara. Melanoidins may have been formed during sun drying of the raw material for 21 days, as occurs during the dehydration of figs, dates and raisins under similar conditions to those employed in cascara processing [11]. In addition, beverages absorbed over the full wavelength range of 200–700 nm, which is characteristic of melanoidins [5]. CGA and melanoidins, usually have a maximum absorption close to 360 nm, which might indicate the linkage of CGA molecules to the structure of melanoidins by non-covalent interactions [5,67,68].

Contribution of MRP to the Antioxidant Properties of IC

In order to identify which compounds were responsible for the antioxidant capacity of IC, the three beverages were ultra-filtrated by a 10 kDa cut-off membrane. Table 4 shows the overall antioxidant capacity of the high (HMW) and low (LMW) molecular weight fractions of the three beverages measured by ABTS and FRAP. Antioxidant capacity analyzed by ABTS and FRAP of the LMW fraction of the three beverages did not differ significantly ($p > 0.05$) among samples. The HMW fractions of the three beverages was significantly ($p < 0.05$) more antioxidant compared to the LMW fraction. The HMW fraction of IC at 10 mg/mL presented the highest values of antioxidant capacity analyzed by ABTS (82.85 mq eq. CGA/mL) and FRAP 1.08 mg eq. CGA/mL). For both antioxidant capacity determinations, the HMW fraction of Tabifruit presented the lowest values. Results seem to indicate that HMW compounds (>10 kDa), such as melanoidins seem to be the main contributors of the overall antioxidant capacity of IC.

Table 4. Antioxidant capacity analyzed by ABTS (mg eq. CGA/mL) and FRAP (mg eq. CGA/mL) of the high molecular weight (HMW, >10 kDa) and low molecular weight (LMW, <10 kDa) fractions of beverages IC (4 mg/mL), IC (10 mg/mL) and Tabifruit separated by ultrafiltration.

	IC (4 mg/mL)		IC (10 mg/mL)		Tabifruit	
	HMW	LMW	HMW	LMW	HMW	LMW
ABTS	61.25 ± 0.49 [b]	0.92 ± 0.01 [a]	82.85 ± 0.19 [d]	0.92 ± 0.00 [a]	66.68 ± 0.48 [c]	0.93 ± 0.00 [a]
FRAP	0.61 ± 0.08 [b]	0.12 ± 0.00 [a]	1.08 ± 0.02 [d]	0.25 ± 0.01 [a]	0.90 ± 0.09 [c]	0.14 ± 0.00 [a]

Values indicate the mean ± standard deviation and different superscript letters denote significant differences between each row (Tukey Test. $p < 0.05$).

Melanoidins are high molecular weight, brown-colored and nitrogen-containing compounds generated in the late stages of the Maillard reaction [4]. Rufián-Henares and Pastoriza stated that melanoidins are great contributors to the overall antioxidant intake in the Spanish diet. In addition, these melanoidins come mainly from coffee, followed by biscuits, beer and chocolate [3,69]. Many health-promoting properties are attributed to melanoidins, such as antioxidant, antimicrobial, anti-inflammatory, antihypertensive or prebiotic activity [4]. Melanoidins have also been isolated from other coffee by-products, such as silverskin, and have shown antioxidant properties in vitro and a dietary fiber effect in vivo [67]. Melanoidins extracted from spent coffee grounds, the last coffee by-product generated during the beverage elaboration, have also shown antioxidant capacity in vitro [70]. IC may be another sustainable source of melanoidins that would contribute to the antioxidant intake of the global population.

3.3. Shelf Life Study Under Accelerated Storage Conditions of Liquid IC

For the stability study, the three liquid beverages were exposed to different conditions of light and temperature for 72 h. Figure 1 shows the differences in color, pH, total antioxidant capacity (TAC) and total phenolic content (TPC) in each beverage exposed to different conditions of light and temperature. As for color, significant differences ($p < 0.05$) were found in L a* b* parameters when the three beverages were exposed to 40 °C and light for 72 h (Table S1). Color alteration after temperature al light exposure can be easily visualized in Figure 1A. Tabifruit was the only sample that presented a significant change ($p < 0.05$) to light treatment, which might suggest that CA has a higher susceptibility to light exposure (Table S1). In general, there is a significant decrease ($p < 0.05$) in the red tonality (a*) of all beverages after temperature and light exposure (Table S1).

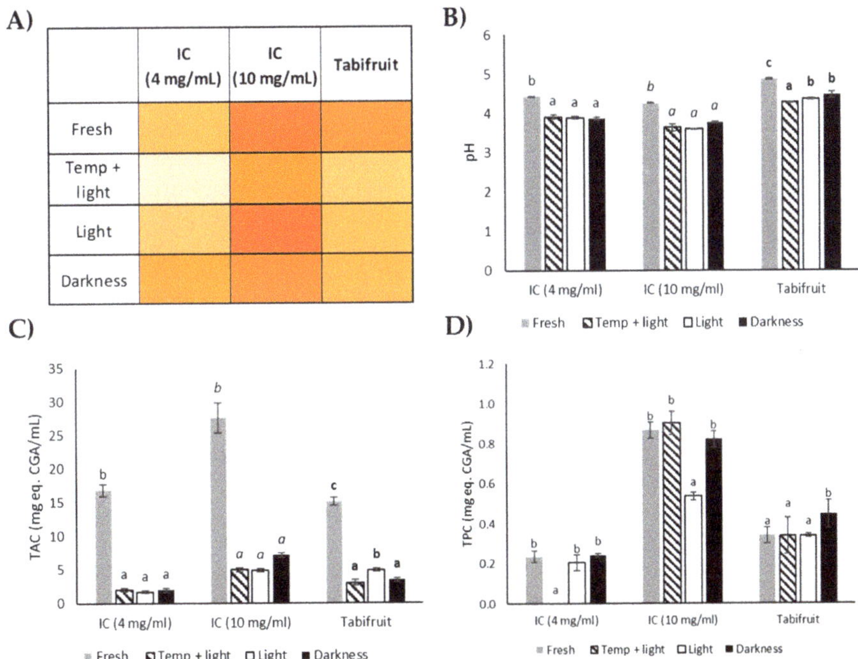

Figure 1. Photosensitivity and heat resistance of IC at 4 mg/mL, IC at 10 mg/mL and Tabifruit. (**A**) Color differences (colorizer.org as a tool for color generation [34]); (**B**) pH; (**C**) Total antioxidant capacity (TAC) determined by ABTS; (**D**) Total phenolic content (TPC). Bars represent the mean values and the error bars denote the standard deviation. Different letters on bars indicate significant differences in each beverage sample (Tukey Test, $p < 0.05$).

For IC at 4 and 10 mg/mL, parameters L and a* correspond linearly to IC concentration (Table S1). Torres et al. (2019) studied the color parameters in coffee cascara before and after drying process. They observed a darker coloration in parameter a* (red color) in the dry samples, concluding that it may be related to the browning processes as effects of the temperature used in the drying of the cascara [71]. Altogether, melanoidins generated during the drying process of coffee cascara seem to be responsible for the color of IC and Tabifruit. This is in accordance to the UV-Visible spectra of the beverages (Figure S1A), which absorb throughout the whole spectrum in a similar way to the caramel standard used to determine melanoidin content (Figure S1B) [72].

Regarding pH results, pH values of all beverages significantly decreased ($p < 0.05$) after the stability study compared to the fresh preparation of the beverages (Figure 1B). These results seem to indicate that temperature and light are determining factors for changes in pH. Nicoli et al. explain that a drop in pH may occur as a consequence of reactions that might be related to non-enzymatic browning, the Maillard reaction, between carbohydrates and amino acids that occurs also during storage [73,74]. The pH is a determining factor for flavor, color, and shelf life of flavored beverages, and it is estimated that the optimal pH to prevent the growth of bacteria and accentuate flavor notes is between 3 and 4 [75]. Other authors state that drinks with acidic pH show good color stability under refrigerated storage conditions to maintain their phenolic content at 90% for a period of 4 months approximately [76]. Therefore, the low pH of IC is suitable for the preservation of the bioactive compounds present in the beverage.

Figure 1C shows results for TAC of the beverages after the stability study. A significant decrease ($p < 0.05$) in antioxidant capacity was observed in all samples compared to the fresh preparation. Considering the TPC (Figure 1D), temperature and light and only light exposure produced a significant decrease in TPC in IC at 4 and 10 mg/mL, respectively. Keeping the liquid preparation in darkness seemed to be the condition that best preserved the phenolic compounds present in IC and Tabifruit.

Considering the composition in nutrients and antioxidants, together with the stability analysis, powdered IC could be considered as the best option for its future commercialization. The process of obtaining IC is green, simple and could be easily carried out using the same facilities used to produce instant coffee. A stability study can predict and optimize the most suitable package conditions to conserve the beverage [33]. Taking into account results from the stability study, IC could be packaged as an instant beverage format to dissolve in either warm or cold water at 10 mg/mL, a similar dosage as the one used for commercially available instant coffee products. Powdered IC would have longer shelf life and less distribution costs, but less expensive concentration and drying techniques, such as vacuum drying at low temperatures (50 °C) and for short periods [77], are needed to implement this procedure in coffee producing countries.

4. Conclusions

A safe instant beverage with antioxidant properties has been obtained to which the following nutrition claims can be assigned: "low fat", "low sugar", "high in fiber" and "source of potassium, magnesium and vitamin C". For the first time, cascara beverage color was attributed to the presence of melanoidins. The shelf life study seemed to indicate that IC beverages in solution are more susceptible to color changes by light and temperature (40 °C) exposure. A package is therefore suggested that protects the product from light and is stored in a cool, dry place. Although very low levels of acrylamide were reported, melanoidins with potential health-promoting properties have also been formed. These melanoidins' high molecular weight compounds (>10 kDa) seem to contribute mostly to the overall antioxidant capacity of IC. The novel powdered instant beverage developed in the present study, IC, is a potential sustainable alternative for instant coffee, with low caffeine and acrylamide levels and a healthy composition in nutrients and antioxidants that would allow the whole recovery of the by-product in two novel ingredients (IC and a dietary fiber fraction).

Supplementary Materials: The following are available online at http://www.mdpi.com/2304-8158/9/5/620/s1, Figure S1: (A) UV-Visible absorption spectrum of the three beverages, measured in a wavelength range of 200 to 700 nm. (B) UV-Visible spectrum of caramel E-150d (Car) and cyanidine-3-glucoside (C-3) at pH 1 and 4.5; Table S1: Measurement of color (L, a*, b*) of the three beverages IC (4 mg/mL), IC (10 mg/mL) and Tabifruit after stability test for 72 h. Studied conditions were: Fresh, 40 °C and light (Temp + light), room temperature and light (light) and room temperature and no light exposure (darkness).

Author Contributions: Conceptualization and investigation, M.D.d.C.; Methodology, J.A.M., E.I. and M.D.d.C.; Data Curation, A.I.-D., A.S.E., M.B.R., M.I.-D., and R.M.; Formal Analysis, A.I.-D., and M.D.d.C.; Writing—Original Draft Preparation, A.S.E., A.I.-D. and M.D.d.C.; Writing—Review & Editing, A.I.-D., M.I.-D., J.A.M., E.I., R.M. and M.D.d.C.; Supervision, A.I.-D. and M.D.d.C.; Project Administration, M.D.d.C.; Funding Acquisition, J.A.M and M.D.d.C. All authors have read and agreed to the published version of the manuscript.

Funding: The project "Generar oportunidades de desarrollo sostenible para 990 familias de La Paz (Honduras) mediante el aprovechamiento de subproductos del café" funded by AECID (2018/ACDE/000666), "Nuevos conocimientos para la sostenibilidad del sector cafetero" funded by CSIC (201970E117) and "Generación de nuevos ingredientes y alimentos beneficiosos dirigidos a condiciones de riesgo y al bienestar global de personas con cáncer colorrectal (TERÁTROFO)" funded by CDTI (IDI-20190960), funded this work.

Acknowledgments: The authors thank the Analysis Service Unit facilities of ICTAN and the Servicio de Química de Proteínas of CIB for the analysis of fatty acids and the amino acid profile. We are grateful to COPADE, COMSA and COMBRIFOL staff for their great contribution to the project. The authors also thank Supracafé for providing Tabifruit and coffee cascara for IC elaboration.

Conflicts of Interest: The authors declare no conflict of interest.

References

1. Cagliero, C.; Ho, T.D.; Zhang, C.; Bicchi, C.; Anderson, J.L. Determination of acrylamide in brewed coffee and coffee powder using polymeric ionic liquid-based sorbent coatings in solid-phase microextraction coupled to gas chromatography-mass spectrometry. *J. Chromatogr. A* **2016**, *1449*, 2–7. [CrossRef]

2. Tamanna, N.; Mahmood, N. Food Processing and Maillard Reaction Products: Effect on Human Health and Nutrition. *Int. J. Food Sci.* **2015**, *2015*, 526762. [CrossRef] [PubMed]

3. Fogliano, V.; Morales, F.J. Estimation of dietary intake of melanoidins from coffee and bread. *Food Funct.* **2011**, *2*, 117–123. [CrossRef] [PubMed]

4. Mesías, M.; Delgado-Andrade, C. Melanoidins as a potential functional food ingredient. *Curr. Opin. Food Sci.* **2017**, *14*, 37–42. [CrossRef]

5. Iriondo-dehond, A.; Ramírez, B.; Escobar, F.V. Antioxidant properties of high molecular weight compounds from coffee roasting and brewing byproducts. *Bioact. Compd. Health Dis.* **2019**, *2*, 48–63. [CrossRef]

6. Soares, C.M.D.; Alves, R.C.; Oliveira, M.B.P.P. Acrylamide in coffee. In *Processing and Impact on Active Components in Food*; Elsevier: Amsterdam, The Netherlands, 2015; pp. 575–582. ISBN 9780124047099.

7. European Food Safety Authority Scientific Opinion on acrylamide in food. *EFSA J.* **2015**, *13*, 4104.

8. Pedreschi, F.; Mariotti, M.S.; Granby, K. Current issues in dietary acrylamide: Formation, mitigation and risk assessment. *J. Sci. Food Agric.* **2014**, *94*, 9–20. [CrossRef]

9. Esquivel, P.; Jiménez, V.M. Functional properties of coffee and coffee by-products. *Food Res. Int.* **2012**, *46*, 488–495. [CrossRef]

10. Rios, M.B.; Iriondo-deHond, A.; Iriondo-deHond, M.; Herrera, T.; Velasco, D.; Gomez-Alonso, S.; Callejo, M.J.; del Castillo, M.D. Effect of coffee cascara dietary fiber on the physicochemical, nutritional and sensory properties of a gluten-free bread formulation. *Molecules* **2020**, *25*, 1358. [CrossRef]

11. Sanz, M.L.; del Castillo, M.D.; Corzo, N.; Olano, A. Formation of amadori compounds in dehydrated fruits. *J. Agric. Food Chem.* **2001**, *49*, 5228–5231. [CrossRef]

12. European Commission COMMISSION REGULATION (EU) 2017/2158 of 20 November 2017 establishing mitigation measures and benchmark levels for the reduction of the presence of acrylamide in food. *Off. J. Eur. Union* **2017**, *204*, 24–44.

13. Martínez-Saez, N.; Del Castillo, M.D. Development of Sustainable Novel Foods and Beverages Based on Coffee By-Products for Chronic Diseases. *Encycl. Food Secur. Sustain.* **2018**, 307–315. [CrossRef]

14. Alves, R.C.; Rodrigues, F.; Nunes, M.A.A.; Vinha, A.F.; Oliveira, M.B.P.P. State of the art in coffee processing by-products. In *Handbook of Coffee Processing By-Products: Sustainable Applications*; Galanakis, C., Ed.; Academic Press-Elsevier: Burlington, MA, USA, 2017; pp. 1–26. ISBN 978-0-12-811290-8.

15. Murthy, P.S.; Madhava Naidu, M. Sustainable management of coffee industry by-products and value addition—A review. *Resour. Conserv. Recycl.* **2012**, *66*, 45–58. [CrossRef]

16. The European Parliament and The Council of the European Union Regulation (EU) 2015/2283 on novel foods. *Off. J. Eur. Union* **2015**, *327*, 1–22.

17. EFSA Panel on Dietetic Products Guidance on the preparation and presentation of an application for authorisation of a novel food in the context of Regulation (EU) 2015/2283. *EFSA J.* **2016**, *14*, 11.

18. Alaska Health Fair, I. How Much Sugar Is There in a Glass of a Powdered Mix Drink? Available online: https://alaskahealthfair.org/much-sugar-glass-powdered-mix-drink/ (accessed on 23 March 2020).

19. Tahmassebi, J.F.; BaniHani, A. Impact of soft drinks to health and economy: A critical review. *Eur. Arch. Paediatr. Dent.* **2020**, *21*, 109–117. [CrossRef]

20. del Castillo, M.D.; Ibañez, M.E.; Amigo, M.; Herrero, M.; Plaza del Moral, M.; Ullate, M. Application of Products of Coffee Silverskin in Anti-Ageing Cosmetics and Functional Food. WO 2013/004873 2013. Available online: https://digital.csic.es/handle/10261/92745 (accessed on 30 March 2020).

21. Martinez-Saez, N.; Tamargo García, A.; Domínguez Pérez, I.; Rebollo-Hernanz, M.; Mesías, M.; Morales, F.J.; Martín-Cabrejas, M.A.; del Castillo, M.D. Use of spent coffee grounds as food ingredient in bakery products. *Food Chem.* **2017**, *216*, 114–122. [CrossRef] [PubMed]

22. Sukhija, P.S.; Palmquist, D.L. Rapid method for determination of total fatty acid content and composition of feedstuffs and feces. *J. Agric. Food Chem.* **1988**, *36*, 1202–1206. [CrossRef]

23. Iriondo-DeHond, A.; Cornejo, F.S.; Fernandez-Gomez, B.; Vera, G.; Guisantes-Batan, E.; Alonso, S.G.; Andres, M.I.S.; Sanchez-Fortun, S.; Lopez-Gomez, L.; Uranga, J.A.; et al. Bioaccesibility, metabolism, and excretion of lipids composing spent coffee grounds. *Nutrients* **2019**, *11*, 1411. [CrossRef] [PubMed]

24. Iriondo-DeHond, A.; Rios, M.B.; Herrera, T.; Rodriguez-Bertos, A.; Nuñez, F.; San Andres, M.I.; Sanchez-Fortun, S.; del Castillo, M.D. Coffee silverskin extract: Nutritional value, safety and effect on key biological functions. *Nutrients* **2019**, *11*, 2693. [CrossRef] [PubMed]

25. Masuko, T.; Minami, A.; Iwasaki, N.; Majima, T.; Nishimura, S.-I.; Lee, Y.C. Carbohydrate analysis by a phenol-sulfuric acid method in microplate format. *Anal. Biochem.* **2005**, *339*, 69–72. [CrossRef] [PubMed]

26. *Comite Europeen de Normalisation Food Analysis—Determination of Acrylamide in Food by Liquid Chromatography Tandem Mass Spectrometry (LC-ESI-MS/MS)*; Comite Europeen de Normalisation: Brussels, Belgium, 2015; EN 166182015.

27. Mastovska, K.; Lehotay, S.J. Rapid Sample Preparation Method for LC–MS/MS or GC–MS Analysis of Acrylamide in Various Food Matrices. *J. Agric. Food Chem.* **2006**, *54*, 7001–7008. [CrossRef]

28. Contini, M.; Baccelloni, S.; Massantini, R.; Anelli, G. Extraction of natural antioxidants from hazelnut (Corylus avellana L.) shell and skin wastes by long maceration at room temperature. *Food Chem.* **2008**, *110*, 659–669. [CrossRef]

29. Wrolstad, R.E.; Giusti, M.M. Characterization and Measurement of Anthocyanins by UV-Visible Spectroscopy. *Curr. Protoc. Food Anal. Chem.* **2001**, F1.2.1–F1.2.13. [CrossRef]

30. Tsao, R.; Yang, R.; Young, C.; Zhu, H. Polyphenolic profiles in eight apple cultivars using high-performance liquid chromatography (HPLC). *J. Agric. Food Chem.* **2003**, *51*, 6347–6353. [CrossRef]

31. Oki, T.; Nagai, S.; Yoshinaga, M.; Nishiba, Y.; Suda, I. Contribution of b-Carotene to Radical Scavenging Capacity Varies among Orange-fleshed Sweet Potato Cultivars. *Food Sci. Technol. Res* **2006**, *12*, 156–160. [CrossRef]

32. Benzie, I.F.F.; Strain, J.J. The ferric reducing ability of plasma (frap) as a measure of '" antioxidant power "': The FRAP assay. *Anal. Biochem.* **1996**, *76*, 70–76. [CrossRef]

33. Manzocco, L.; Kravina, G.; Calligaris, S.; Nicoli, M.C. Shelf life modeling of photosensitive food: The case of colored beverages. *J. Agric. Food Chem.* **2008**, *56*, 5158–5164. [CrossRef]

34. Sebastian Loncar Colorizer—Color Picker and Converter. Available online: http://colorizer.org/ (accessed on 20 March 2020).

35. Adrianzén, G. Determinación De La Capacidad Antioxidante y Polifenoles Totales de la Cáscara Y Mucílago De La Especie Coffea Arábica L. Y Sus Posibles Usos. Bachelor's Thesis, Universidad Nacional José Faustino Sánchez Carrión, Huacho, Peru, 2018.

36. Alvarez, W. Evaluación De Rangos Del Espectro Caracterización Fisicoquímica De Cáscara De Café Para Alimentación Animal. Master's Thesis, Universidad Nacional Toribio Rodríguez Mendoza de Amazonas, Chachapoyas, Peru, 2018.

37. Tobón Arroyave, N.; Cerón Cárdenas, A.F.; Garcés Giraldo, L.F. Análisis y modelamiento de la granulometría en la cáscara del café (Coffea arabica L.) variedad Castillo. *Producción+ Limpia* **2015**, *10*, 80–91. [CrossRef]

38. Pandey, A.; Soccol, C.R.; Nigam, P.; Brand, D.; Mohan, R.; Roussos, S. Biotechnological potential of coffee pulp and coffee husk for bioprocesses. *Biochem. Eng. J.* **2000**, *6*, 153–162. [CrossRef]

39. Elías, L.G. Chemical composition of coffee-berry by-products. *Coffee Pulp Compos. Technol. Util.* **1979**, 11–16.

40. Heeger, A.; Kosińska-Cagnazzo, A.; Cantergiani, E.; Andlauer, W. Bioactives of coffee cherry pulp and its utilisation for production of Cascara beverage. *Food Chem.* **2016**, *221*, 969–975. [CrossRef] [PubMed]

41. Janissen, B.; Huynh, T. Chemical composition and value-adding applications of coffee industry by-products: A review. *Resour. Conserv. Recycl.* **2018**, *128*, 110–117. [CrossRef]

42. World Health Organization. Joint WHO-FAO expert consultation on diet, nutrition, and the prevention of chronic diseases, 2002, Geneva, Switzerland. *WHO Tech. Rep. Ser.* **2003**, *916*, 149.

43. Urbaneja, G.; Ferrer, J.; Paez, G.; Arenas, L.; Colina, G. Acid hydrolysis and carbohydrates characterization of coffee pulp. *Renew. Energy* **1996**, *9*, 1041–1044. [CrossRef]

44. Belliveau, D. The Coffee Cherry Co. Available online: https://coffeecherryco.com/ (accessed on 24 March 2020).

45. Bernacchia, R.; Preti, R.; Vinci, G. Organic and conventional foods: Differences in nutrients. *Ital. J. Food Sci.* **2016**, *28*, 565–578.

46. Ito, H.; Ueno, H.; Kikuzaki, H. Nutrition and dietetic practice free amino acid compositions for fruits. *J. Nutr. Diet. Pr.* **2017**, *1*, 1–5.

47. Takaoka, M.; Okumura, S.; Seki, T.; Ohtani, M. Effect of amino-acid intake on physical conditions and skin state: A randomized, double-blind, placebo-controlled, crossover trial. *J. Clin. Biochem. Nutr.* **2019**, *65*, 52–58. [CrossRef]

48. Turck, D.; Castenmiller, J.; De Henauw, S.; Hirsch-Ernst, K.I.; Kearney, J.; Knutsen, H.K.; Maciuk, A.; Mangelsdorf, I.; McArdle, H.J.; Naska, A.; et al. Guidance on the scientific requirements for health claims related to muscle function and physical performance: (Revision 1). *EFSA J.* **2018**, *16*.

49. Regulation EC 1924 European Community (EC) No 1924/2006 of the European Parliament and of the Council of 20 December 2006 on nutrition and health claims made on foods. *Off. J. Eur. Union* **2006**, 9–25.

50. Chawla, R.; Patil, G.R. Soluble dietary fiber. *Compr. Rev. Food Sci. Food Saf.* **2010**, *9*, 178–196. [CrossRef]

51. Koh, A.; De Vadder, F.; Kovatcheva-Datchary, P.; Bäckhed, F. From dietary fiber to host physiology: Short-chain fatty acids as key bacterial metabolites. *Cell* **2016**, *165*, 1332–1345. [CrossRef] [PubMed]

52. The european commission commission regulation (EU) No 432/2012. *Off. J. Eur. Union* **2012**, *13*.

53. European council council directive of 24 September 1990 on nutrition labelling for foodstuffs (90/496/ EEC). *Off. J. Eur. Union* **1990**, 18–63.

54. Hasanah, U.; Setyowati, M.; Efendi, R.; Safitri, E.; Idroes, R.; Heng, L.Y.; Sani, N.D. Isolation of Pectin from coffee pulp Arabica Gayo for the development of matrices membrane. *IOP Conf. Ser. Mater. Sci. Eng.* **2019**, *523*, 12–14. [CrossRef]

55. USDA (United States Department of Agriculture USA) FoodData Central. Available online: https://fdc.nal.usda.gov/fdc-app.html#/?query=hemp (accessed on 25 March 2020).

56. Porto, A.C.V.; Freitas-Silva, O.; de Souza, E.F.; Gottschalk, L.M.F. Effect of asparaginase enzyme in the reduction of asparagine in green coffee. *Beverages* **2019**, *5*, 32. [CrossRef]

57. Onishi, Y.; Prihanto, A.A.; Yano, S.; Takagi, K.; Umekawa, M.; Wakayama, M. Effective treatment for suppression of acrylamide formation in fried potato chips using L-asparaginase from Bacillus subtilis. *3 Biotech* **2015**, *5*, 783–789. [CrossRef]

58. Atanda, S. A Fungi and mycotoxins in stored foods. *Afr. J. Microbiol. Res.* **2011**, *5*, 4373–4382. [CrossRef]

59. Iriondo-DeHond, A.; Aparicio García, N.; Velazquez Escobar, F.; San Andres, M.I.; Sanchez-Fortun, S.; Blanch, G.P.; Fernandez-Gomez, B.; Guisantes Batan, E.; del Castillo, M.D. Validation of coffee by-products as novel food ingredients. *Innov. Food Sci. Emerg. Technol.* **2019**, *51*, 194–204. [CrossRef]

60. European Food Safety Authority (EFSA) Scientific Opinion on the safety of caffeine. *EFSA J.* **2015**, *13*, 1–21.

61. del Castillo, M.D.; Iriondo-DeHond, A.; Martinez-Saez, N.; Fernandez-Gomez, B.; Iriondo-DeHond, M.; Zhou, J.-R. Chapter 6—Applications of recovered compounds in food products. In *Handbook of Coffee Processing By-Products*; Academic Press-Elsevier: London, UK, 2017; pp. 171–194. ISBN 9780128112908.

62. Murthy, P.S.; Naidu, M.M. Recovery of phenolic antioxidants and functional compounds from coffee industry by-products. *Food Bioprocess Technol.* **2012**, *5*, 897–903. [CrossRef]

63. Karadeniz, F.; Durst, R.W.; Wrolstad, R.E. Polyphenolic composition of raisins. *J. Agric. Food Chem.* **2000**, *48*, 5343–5350. [CrossRef]

64. Murthy, P.S.; Manjunatha, M.R.; Sulochannama, G.; Naidu, M.M. Characterization and bioactivity of coffee anthocyanins. *Eur. J. Biol. Sci.* **2012**, *4*, 13–19.

65. Prata, E.R.; Oliveira, L.S. Fresh coffee husks as potential sources of anthocyanins. *LWT Food Sci. Technol.* **2007**, *40*, 1555–1560. [CrossRef]

66. Belay, A.; Gholap, A. Characterization and determination of chlorogenic acids (CGA) in coffee beans by UV-Vis spectroscopy. *Afr. J. Pure Appl. Chem.* **2009**, *3*, 234–240.

67. Tores de la Cruz, S.; Iriondo-DeHond, A.; Herrera, T.; Lopez-Tofiño, Y.; Galvez-Robleño, C.; Prodanov, M.; Velazquez-Escobar, F.; Abalo, R.; del Castillo, M.D. An assessment of the bioactivity of coffee silverskin melanoidins. *Foods* **2019**, *8*, 68. [CrossRef]

68. Delgado-Andrade, C.; Rufián-Henares, J.A.; Morales, F.J. Assessing the antioxidant activity of melanoidins from coffee brews by different antioxidant methods. *J. Agric. Food Chem.* **2005**, *53*, 7832–7836. [CrossRef]

69. Pastoriza, S.; Rufián-Henares, J.A. Contribution of melanoidins to the antioxidant capacity of the Spanish diet. *Food Chem.* **2014**, *164*, 438–445. [CrossRef]

70. Bravo, J.; Juániz, I.; Monente, C.; Caemmerer, B.; Kroh, L.W.; De Peña, M.P.; Cid, C. Evaluation of spent coffee obtained from the most common coffeemakers as a source of hydrophilic bioactive compounds. *J. Agric. Food Chem.* **2012**, *60*, 12565–12573. [CrossRef]

71. Torres-Valenzuela, L.S.; Martínez, K.G.; Serna-Jimenez, J.A.; Hernández, M.C. Secado de pulpa de café: Condiciones de proceso, modelación matemática y efecto sobre propiedades fisicoquímicas. *Inf. Tecnológica* **2019**, *30*, 189–200. [CrossRef]

72. Royle, L.; Radcliffe, C.M. Analysis of caramels by capillary electrophoresis and ultrafiltration. *J. Sci. Food Agric.* **1999**, *79*, 1709–1714. [CrossRef]

73. Nicoli, M.C.; Calligaris, S.; Manzocco, L. Shelf-life testing of coffee and related products: Uncertainties, pitfalls, and perspectives. *Food Eng. Rev.* **2009**, *1*, 159–168. [CrossRef]

74. Manzocco, L.; Lagazio, C. Coffee brew shelf life modelling by integration of acceptability and quality data. *Food Qual. Prefer.* **2009**, *20*, 24–29. [CrossRef]

75. Malvais, R. Estudio De Vida De Anaquel En Bebidas Saborizadas. Bachelor's Thesis, Universidad Autonóma del Estado de México, Cuernavaca, Mexico, 2017.

76. Marete, E.N.; Jacquier, J.C.; O'Riordan, D. Feverfew as a source of bioactives for functional foods: Storage stability in model beverages. *J. Funct. Foods* **2011**, *3*, 38–43. [CrossRef]

77. Michalska, A.; Wojdyło, A.; Brzezowska, J.; Majerska, J.; Ciska, E. The influence of inulin on the retention of polyphenolic compounds during the drying of blackcurrant juice. *Molecules* **2019**, *24*, 4167. [CrossRef]

MDPI

St. Alban-Anlage 66

4052 Basel

Switzerland

Tel. +41 61 683 77 34

Fax +41 61 302 89 18

www.mdpi.com

Foods Editorial Office

E-mail: foods@mdpi.com

www.mdpi.com/journal/foods